Experience and Prediction

Experience and Prediction

An Analysis of the Foundations and the Structure of Knowledge

BY

HANS REICHENBACH, Ph.D.

WITH A NEW INTRODUCTION BY

ALAN W. RICHARDSON

University of Notre Dame Press

Notre Dame, Indiana

Notre Dame, Indiana 46556
www.undpress.nd.edu

Manufactured in the United States of America

Library of Congress Cataloging-in-Publication Data

Reichenbach, Hans, 1891–1953.
Experience and prediction : an analysis of the foundations and the structure of knowledge / by Hans Reichenbach ; with a new introduction by Alan W. Richardson.
p. cm.
Originally published: [Chicago] : University of Chicago Press, c1938. With new introd.
Includes bibliographical references and index.
ISBN-13: 978-0-268-04055-0 (pbk. : alk. paper)
ISBN-10: 0-268-04055-9 (pbk. : alk. paper)
1. Knowledge, Theory of. 2. Experience. I. Title.
BD161.R38 2006
121—dc22
2006005009

∞ *This book is printed on acid-free paper.*

CONTENTS

Introduction — vii
  Alan W. Richardson

Preface — xli

CHAPTER — PAGE

I. MEANING
- § 1. The three tasks of epistemology — 3
- § 2. Language — 16
- § 3. The three predicates of propositions — 19
- § 4. The language of chess as an example, and the two principles of the truth theory of meaning — 28
- § 5. Extension of the physical theory of truth to observation propositions of ordinary language — 33
- § 6. Extension of the truth theory of meaning to observation propositions of ordinary language — 37
- § 7. The meaning of indirect propositions, and the two principles of the probability theory of meaning — 46
- § 8. Discussion of the verifiability theory of meaning — 57

II. IMPRESSIONS AND THE EXTERNAL WORLD
- § 9. The problem of absolute verifiability of observation propositions — 83
- § 10. Impressions and the problem of existence — 88
- § 11. The existence of abstracta — 93
- § 12. The positivistic construction of the world — 100
- § 13. Reduction and projection — 105
- § 14. A cubical world as a model of inferences to unobservable things — 114
- § 15. Projection as the relation between physical things and impressions — 129
- § 16. An egocentric language — 135
- § 17. Positivism and realism as a problem of language — 145
- § 18. The functional conception of meaning — 156

III. An Inquiry concerning Impressions
§ 19. Do we observe impressions? 163
§ 20. The weight of impression propositions 169
§ 21. Further reduction of basic statements 179
§ 22. Weight as the sole predicate of propositions 187

IV. The Projective Construction of the World on the Concreta Basis
§ 23. The grammar of the word "existence" 195
§ 24. The different kinds of existence 198
§ 25. The projective construction of the world 203
§ 26. Psychology 225
§ 27. The so-called incomparability of the psychical experiences of different persons 248
§ 28. What is the ego? 258
§ 29. The four bases of epistemological construction 262
§ 30. The system of weights co-ordinated to the construction of the world 273
§ 31. The transition from immediately observed things to reports 282

V. Probability and Induction
§ 32. The two forms of the concept of probability 297
§ 33. Disparity conception or identity conception? 302
§ 34. The concept of weight 312
§ 35. Probability logic 319
§ 36. The two ways of transforming probability logic into two-valued logic 326
§ 37. The aprioristic and the formalistic conception of logic 334
§ 38. The problem of induction 339
§ 39. The justification of the principle of induction 348
§ 40. Two objections against our justification of induction 357
§ 41. Concatenated inductions 363
§ 42. The two kinds of simplicity 373
§ 43. The probability structure of knowledge 387

Index 407

INTRODUCTION

Alan W. Richardson

Hans Reichenbach's *Experience and Prediction* was first published in 1938 by the University of Chicago Press. Reichenbach wrote it during his time at the University of Istanbul, where he fled after being removed from his post at the University of Berlin by the Nazi race laws in 1933. Although it was not the first book of Reichenbach's to appear in English—his series of popular lectures, *Atom and Cosmos*, had been translated and published by George Braziller in 1932—it was the first book Reichenbach wrote in English. The book served self-consciously, therefore, as Reichenbach's entry into the English-speaking philosophical world. Indeed, by the time it appeared in 1938, Reichenbach had already accepted a position at the University of California at Los Angeles, as several of the early reviewers remarked.[1]

Reichenbach was already a well-known figure in the North American philosophical world, a prominent member of a group of European philosophers who promoted the project of logical empiricism.[2] This

I thank Don Howard and Maria Reichenbach for the opportunity to write this introduction. I am grateful to Mike Waters for gathering much of the early literature the book generated, discussing the book with me as I drafted this introduction, and for comments on an earlier draft. I am also grateful for detailed comments on an earlier draft from Don Howard, Elliott Sober, and Thomas E. Uebel. I thank the Archives of Scientific Philosophy in the Twentieth Century at the University of Pittsburgh for allowing me to quote from several documents in the Hans Reichenbach Collection (cited using the Archives' numbering system: HR nnn-nn-nn).

1. The University of Chicago Press accepted the manuscript in early 1937, thereby confirming the evidence in the text (section 21) that the main writing was done in 1936 (Donald Bean [?] to Hans Reichenbach, 2 April 1937; HR 040-21-18).

2. Reichenbach had already, for example, written a substantial essay for the *Journal of Philosophy* on the German contribution to logical empiricism (Reichenbach 1936), a contribution that led him to prefer the term "logical empiricism" to describe his own philosophy to the name that Herbert Feigl and Alfred Blumberg (1931) used to name the philosophy of the Vienna Circle: "logical positivism."

fact led to the book being both prominently reviewed and reviewed as a contribution to logical empiricism[3] Moreover, Reichenbach is quite clear in his preface that the book ought to be read as a contribution to an international and collaborative project of logical empiricism (he calls it "logistic empiricism" in the preface). Reichenbach's quite prominent divergences from theses considered by some to be definitive of logical empiricism, however, led commentators (e.g., Barrett 1939) to understand the book to be a sign that there was a crisis among the logical empiricists—a crisis evident in the fact that the most negative review and most critical early commentary on the book were authored by the young American philosopher who had come to be regarded by many as a recruit to the project, Ernest Nagel (1938, 1939).

Viewed from an historical distance, however, *Experience and Prediction* came to be understood not as a sign of a crisis within logical empiricism, but as a significant document in founding the North American interpretation of a proper logical empiricist project. The book stands with other work of the mid to late 1930s, prominently including Rudolf Carnap's *Logical Syntax of Language* (Carnap [1934] 1937) and "Testability and Meaning" (Carnap 1936/1937), that promoted a liberalized logical empiricism not committed to strict verificationism or the reduction of all scientific discourse into a language of sensation.[4] Indeed, Reichenbach's remarks in the first section of *Experience and Prediction* on the task of a philosophy of scientific knowledge form perhaps the single most influential discussion of the proper goals of logical empiricist philosophy of science. Arguably, the long-term significance of *Experience and Prediction* has been as the single most compelling account of a project, a method, and a line of research that came to be the canonical account of logical empiricism.

3. The book was reviewed in the leading philosophy journals: in *Journal of Philosophy* by Ernest Nagel (1938), in *Philosophical Review* by William R. Dennes (1939), in *Philosophy of Science* by Eleanor Bisbee (1938), in *Ethics* by W. H. Werkmeister (1938). Reviews also appeared, for example, in *The Nation* (Hook 1939) and *American Journal of Sociology* (House 1939).

4. The earliest and most vocal supporter of physicalism among the logical empiricists was Otto Neurath, of course, but this fact was not widely known among American readers of Reichenbach in the 1930s, except through Carnap's citations of his work.

Experience and Prediction: Its Relation to Reichenbach's Earlier Work

Hans Reichenbach (1891–1953) is an important figure in the development of logical empiricist philosophy of science, with one foot in two superficially very different sorts of literature.[5] On the one hand, he was a leading figure in several of the most technical areas of logical empiricist work. He was the most important philosopher of physics in the first generation of logical empiricism, doing fundamental work in the foundations of space-time theories, in quantum mechanics, in issues in the direction of time, and in statistical mechanics. He also was a leading figure in the development of the logical foundations of probability theory. At the same time, however, he was much more diligent than most of the logical empiricists in popularizing both science and philosophy of science. From his early book, *Atom and Cosmos*, which treated of the new physics' lessons for understanding the largest and smallest scales of existence, through his late (1951) tour de force and academic bestseller, *The Rise of Scientific Philosophy*, he showed a willingness rare among philosophers of the time to speak of the larger themes of philosophy of science to an educated lay audience. *Experience and Prediction* is neither a technical work of philosophy of science nor a popular book. It is an attempt by Reichenbach, occasioned by his desire to move into a new philosophical context in North America, to explain to other professional philosophers the epistemological project that the technical work both suggested and served.[6]

Reichenbach's attempt at a less technical exposition of a general epistemological point of view was surprising to some of the readers of the book who were familiar with his early writings. In 1938 Reichenbach was already quite well known among those American philosophers who could read German for his important work in the foundations of space-time theories. In a series of books in the 1920s, Reichenbach worked

5. A brief biography and fulsome bibliography of Reichenbach can be found in Stadler (1997, 902–10). See also the biography in Reichenbach (1978, 413–23).

6. It is quite clear in editorial correspondence with Charles Morris that Reichenbach's book was well into production before the offer of a position from UCLA was in hand. As late as October 1937, Reichenbach's letters to Morris indicate that the UCLA offer was still pending (Hans Reichenbach to Morris, 24 October 1937, HR 040-21-07).

out a conventionalist methodology of relativistic physics, arguing that the metric of space-time relied on conventional definitions of crucial notions such as "simultaneity." Reichenbach, like other young philosophers of the exact sciences of his generation, was deeply impressed by the far-reaching changes in physics brought about by Einstein's special and general theories of relativity. Moreover, like many who reflected on Einstein's accomplishment in the 1910s and 1920s, Reichenbach adopted a conventionalist line on the achievement: According to Reichenbach, what Einstein showed was that concepts like length, duration, and mass depended upon conventional or definitional decisions regarding notions like identity of length of objects at a distance from one another, simultaneity of events at a distance from one another, or identity of mass of two distinct objects.[7] Such definitions first allow a coordination between an independently well-worked-out and purely mathematical metric geometry and physical events and processes. Since such conventions, thus, first allow mathematical description of the world, they are requisites for the mathematical physics that is our best understanding of the physical world. By 1928, with the publication of his *Philosophie der Raum-Zeit-Lehre* (later translated as *Philosophy of Space and Time* [1958]), in which he sought to establish and draw out the consequences of this sort of conventionalism, Reichenbach had established himself as one of the most technically proficient philosophers of physics of his generation. This reputation was augmented by the fact that Einstein himself had intervened to get Reichenbach a position teaching philosophy of science in the physics faculty at the University of Berlin, a post Reichenbach took up in 1926.

Sandwiched around the work in philosophy of space-time physics was Reichenbach's other technical work in the foundations of probability and statistics. This interest was already evident in Reichenbach's PhD dissertation of 1915 at the University of Erlangen, "Der Begriff der Wahrscheinlichkeit für die mathematische Darstellung der Wirklichkeit" (The Concept of Probability in the Mathematical Representation of Reality), published in three parts in the journal *Zeitschrift für Philosophie und*

7. As we shall see below, Reichenbach first discussed these methodological lessons in a robustly Kantian vocabulary, as a relativized a priori.

philosophische Kritik in 1916 and 1917 (Reichenbach 1916/1917). It is also evident in the last important book Reichenbach wrote in German, his 1935 *Wahrscheinlichkeitslehre: Eine Untersuchung über die logischen und mathematischen Grundlagen der Wahrscheinlichkeitsrechnung* (Probability Theory: An Investigation of the Logical and Mathematical Foundations of the Calculus of Probability). The primary arguments of Reichenbach's early work in foundations of probability theory are fourfold. First, he argued that any system of scientific knowledge of the world depended upon probabilistic inference. Second, he attempted to provide an account of the foundations of probability consistent with empiricism, adopting, therefore, a frequency account of probability. Third, he argued that a univocal account of probability was possible; in particular, he sought to give a frequency interpretation of probabilities both for physical events in the world (the probability that this die will come up a two on the next toss, for example) and for logical relations among propositions (the probability that Einstein's general theory of relativity is true, given the truth of Eddington's solar eclipse data, for example). Finally, he sought to establish probability theory itself as a generalization to continuously many truth values of traditional two-valued logic.

Thus, Reichenbach's appearance as a general epistemologist in 1938 was, for some of his readers, a bit puzzling. While the work on probability and induction is front and center in *Experience and Prediction*, his technical work in philosophy of physics is rather submerged. It appears, really, only as a set of solutions to problems that is then used to motivate a general, if tempered, voluntarism in the epistemology of *Experience and Prediction*. Thus, for example, in section 27, the problem of the incomparability of the experiences of any two people is assimilated to the need for coordinative definitions for psychological terms, explicitly following Reichenbach's solution to methodological issues in the philosophy of space and time. Indeed, the conventionalism Reichenbach takes to be established prior to *Experience and Prediction* is the motivation for the single most striking and characteristic part of his epistemology: the stress on the need for volitional decisions that is remarked upon already in section 1.

Since the conventionalism in physics that Reichenbach proffered is presupposed and neither precisely expressed nor argued for in the text and since it plays such a large role in shaping Reichenbach's general theory of knowledge, we would do well to present a simple version of it here.[8] One way into the conventionalism of Reichenbach is simply to think about how one would discover the geometrical structure of space(-time). How would one perform an experiment to determine whether space(-time) is Euclidean or not? Well, on the side of the mathematics, by the late nineteenth century, the differences between Euclidean and various non-Euclidean geometries were well understood. Thus, for example, the Euclidean geometry of three-dimensional space is the unique geometry in which the sum of the angles of a triangle, no matter the size of the triangle, is 180 degrees. In the other, the non-Euclidean, geometries (on "curved" space), the sums of the angles of triangles diverge more and more from 180 degrees as the triangles get larger and larger. Thus, one appears to have a straightforward procedure for discovering the geometry of space: measure the angles of a very large triangle and see what they sum to.

Now, waiving issues of measurement error (which merely renders the results subject to standard probabilistic hedges), Reichenbach reminds us that there is a serious problem with this procedure. While the mathematical facts may be perfectly well understood, no operation will allow us to ascertain them for physical space unless and until we know what a physical line is—triangles, after all, need to have straight sides. Thus, if one sets up the experiment à la Carl Friedrich Gauss, using light beams, one presupposes that light travels in straight lines. Here we seem to have only unsavory options. We might, in an intuitionist empiricist mood, declare that we have a direct experiential knowledge of the straightness of light beams. This confident assertion, however, lacks plausibility and suffers greatly under the pressure of Einstein's theoretical advances, which proceed in large measure by calling into question intuitive claims like this one. We might, in a strict empiricist mood, declare that no empirical sense can be made of physical straightness.

8. An elegant, elementary, but much more precise account of metrical conventionalism can be found in Sklar (1974). There is a large literature on Reichenbach's conventionalism. Three excellent entry points to this literature are Ryckman (1992), Friedman (1999, chapters 3 and 4), and Howard (1994). For a more expansive account of the place of philosophy of physics within the projects of logical empiricism, see Ryckman (forthcoming).

This sort of empiricism ends up dismantling the entire edifice of modern science, however, since if straightness is denied empirical sense, length is next and the whole business of measuring distance, velocity, and so on follows.

Finally, we might try to solve this problem nonempirically, perhaps by declaring the geometry of physical space knowable in a priori ways. This way lies a strict Kantianism, for example, which tries to provide a transcendental explanation of the necessity of precisely Euclidean geometry. Here again we have two problems, one a very general philosophical problem and one a philosophico-historical problem well entrenched by 1928. The general problem is providing a persuasive account of the nonempirical origins of our knowledge of the geometry of space. The more pressing problem, however, is the historical problem mentioned above: by 1928, Einstein has offered new definitions and procedures that seem to give the lie to confident assertions about the a priori origin of knowledge of Euclidean geometry. While no such historical facts render it impossible for there to be a priori knowledge of genuine geometrical structures, philosophers of science were, by Reichenbach's generation, deeply impressed by how independent physical theorizing seemed to be from constraints on knowledge firmly argued for by a priorist philosophies.

Reichenbach followed in a line of conventionalist thinkers, beginning with the French physicist and mathematician, Henri Poincaré ([1902] 1946), who made a virtue of a seeming necessity here.[9] Suppose the only way to make sense of the procedure by which Gauss purported to measure the geometry of space (sending light beams from three mountain tops in order to create a "light triangle" whose angles could be measured) was to say that Gauss presupposed that light traveled along straight lines. Instead of looking for the foundations of Gauss's knowledge that light traveled in straight lines, one could declare that Gauss simply made an informal and tacit decision to treat "path of a light ray" as a physical definition of "straight line." In order to measure length one needs to have some definition of straightness that renders it physically ascertainable. One does not *discover* what constitutes

9. Just how strictly Reichenbach follows Poincaré is one theme in the literature on conventionalism cited above.

physical straightness, one *decides* what will count as straight. Similarly, for example, time measurement needs units of time and a procedure for determining whether two events at a distance from one another occurred at the same time. The conventionalist will say "the same amount of time" (for the unit) and "occurring at the same time" (for distant events) are, or rely on, conventional decisions. Conventions, since they must be presupposed for knowledge claims to be formulated at all (for example, the laws of dynamics rely on concepts whose physical significance presupposes that we are able to measure time and space), are not the sorts of things that have further epistemological foundations—insofar as there are foundations of physical knowledge, they *are* the epistemological foundations. They serve that role by being fundamental acts, decisions, through which alone we can begin investigating the physical world.

As mentioned, Reichenbach, in the book before us, generalizes and tempers the lessons of conventionalism. He tempers the conclusions by drawing out the relations of conventions and knowledge claims: while decision plays a necessary part in knowledge for Reichenbach, decision does not proceed unconstrained. Rather, decisions up front constrain further decisions downstream; these are the "entailed decisions" of section 1. One cannot choose, for example, the path of light rays as straight lines and choose Euclidean geometry as the global geometry of space. Once one chooses to treat light rays as straight lines, the global Euclidean structure of space becomes a matter of empirical investigation. Obversely, if one chooses Euclidean geometry, one might find oneself constrained to say on empirical grounds that light does not follow straight lines through space.

Nonetheless, the general epistemology of *Experience and Prediction* posits decisions right through the whole of the structure of knowledge. It is not just that physical concepts require conventional definitions—even in psychology there are decisions to be made. Moreover, in a way, the epistemology Reichenbach proffers in 1938 radicalizes the conventionalist idea through its notion of "volitional bifurcations" (section 1). Conventionalism can seem a tame doctrine: one must decide whether to weigh objects in pounds or kilograms. This would be a trivial semantical conventionalism with a clear mapping from one value to another, no more epistemologically interesting than the difference between English and German. Even metrical conventions could be seen as different ways of organizing the same facts. Volitional bifurcations are in no way tame; they are, by definition, decisions such that the alternatives lead to different global structures of knowledge. Reichenbach argues (section 1) that the very aim of science involves a decision. This decision, perforce, will alter what it is possible to know since it alters not merely what we understand by "knowing" but also our very knowledge-gathering practices.

The Reception and Influence of *Experience and Prediction*: Four Central Themes

Experience and Prediction is a rich work with many possible and actual consequences for philosophical thought and practice. Four themes are dominant in the critical attention that the book has received and explain its lasting influence on epistemology and philosophy of science. This section briefly comments on each of these four themes, in turn. Given how interconnected they are in the book, disentangling them is done more for convenience in tracking the influence of Reichenbach's work than for illuminating the work itself.

The first theme can be summarized under the heading "Logic, Semantics, and Probability Theory." Although not a technical work by logical empiricist standards, *Experience and Prediction* was clearly intended to offer an epistemological and semantic framework within which the technical features of probability theory were revealed as philosophical in their own right. Moreover, the book played a significant role in putting such technicalities, as well as foundational issues in the interpretation of probabilities, on the philosophical agenda. In the case of Reichenbach's own philosophical goals, the details of probability theory were alluded to in *Experience and Prediction* in support of a project in semantics and logic, a project that forms the basis of his rejection of a strict sensationalist positivism but which found few obvious champions in the years after 1938.

Reichenbach's main project in *Experience and Prediction* is to marshal probability theory as the foundation for semantics. His rejection of verificationism is predicated on the availability of a new foundation of meaningfulness and synonymy offered by probability. His two principles of the probability theory of meaning in section 7 give the essential content of the account he proposes:

> A proposition has meaning if it is possible to determine a weight, i.e., a degree of probability, for the proposition.
>
> Two sentences have the same meaning if they obtain the same weight, or degree of probability, by every possible observation.

Now, if one were to put the fundamental principles of the verificationist theory of meaning, the theory of meaning that Reichenbach associated with Viennese logical empiricism in 1938, into a similar form, they might read:

> A proposition has meaning if it is possible to verify the proposition.
>
> Two sentences have the same meaning if they are verified equally by every possible observation.

When these theories are placed side by side in this manner, one can easily see why Reichenbach would view his theory as a necessary liberalization of the verificationist account. If one simply sees "degree of probability" as "verified to a certain degree" rather than as "conclusively verified," Reichenbach's liberalization just seems to be the counsel of good empiricist sense.

By using probability as the foundation of what claims about the world even mean, Reichenbach underscored the key theme of the book: the lesson of twentieth-century empiricism is that certainty is not possible in our knowledge of the world. The centrality of this epistemological theme is emphasized in the opening of Eleanor Bisbee's lengthy review of *Experience and Prediction*, entitled "A World of Probability," in *Philosophy of Science* in 1938:

> The first half of the twentieth century will go down in history as the era of intellectual problems which steadily accumulated in science through the first

> three decades and, in 1929, were formulated for general reading in such books as John Dewey's "Quest for Certainty" and A.S. Eddington's "The Nature of the Physical World." Dr. Reichenbach's book, "Experience and Prediction" [*sic*] appears in the fourth decade to present an epistemology in which the nature of the physical world as it can be known by man is admittedly an uncertain quest. How a confident book can be written in terms of uncertainty is admirably demonstrated in this work. (Bisbee 1938, 360–61)

Indeed, Reichenbach's book is an epistemologically confident book: the uncertainty of knowledge of the physical world is all-pervasive, yet, far from underwriting skepticism, this very uncertainty is the source of Reichenbach's rejection of a sensationalist positivism and defense of realism. The general idea of the defense of realism is simple: Once we have rejected the strict verificationist doctrine, we must see that claims about the physical world cannot be translated into claims about experience or sensory impressions. Having given up translation reduction in favor of probabilistic inferences from impressions to the external world, we can see that the positing of physical objects that exist independently of our impressions is not semantically equivalent to any claim about our impressions. The next step for Reichenbach is to argue, on a rather underspecified inference using Bayes' Theorem, that systematic interconnections among our impressions are made more probable by the posit of physical objects that cause them than on any positivist thesis that simply posits those interconnections. The final step is to argue from the observed interconnections among our impressions backward to the higher probability of the realist thesis that there are physical objects that cause those interconnections than that those interconnections simply exist as such. This whole argument is given in detail in Reichenbach's charming thought experiment involving the cubical world and systematic interconnections among visible shadows that he provides in chapter 2.

Before taking up the problem of the external world per se, two more points about probability as the first key theme of the book should be made. First, there were dissenters from the very beginning about whether probability, however central it may be to epistemology, could be a proper foundation for semantics. A prominent and vocal dissenter from early on was Nagel, who wrote both a short review and a lengthy response to the book. In both pieces, Nagel took probabilistic semantical theses to task. Consider Reichenbach's synonymy criterion—two sentences have the same meaning if they obtain the same weight, or degree of probability, by every possible observation. Nagel objected that equiprobability could not possibly be a sufficient condition of synonymy:

> To take a trivial example, when we are tossing a fair coin the sentences "This coin will fall heads uppermost next toss" and "This coin will fall tails uppermost on the next toss" are assigned equal weights but distinct meanings. Equality of weights is at best a necessary condition of sameness of meaning. (Nagel 1938, 272)

It is not difficult to find Nagel's objection here unsympathetic and, thus, too quick. Reichenbach's likely response is that Nagel's case is considered too holistically. "This coin is heads" and "this coin is tails" are, we shall suppose, given distinct weights in many epistemic situations. Thus, since "this coin lands heads on the n-th toss" is probabilistically distinguishable from "this coin lands tails on the n-th toss" at least *after* the n-th toss, we in fact do not assign the same weight for these two hypotheses on *every experience.* The chief problem in the Nagel case is the use of the temporal indexical word "next," but we have reason to suppose a properly compositional semantics and tense logic will finesse this problem also. So, the general lines of a Reichenbachian response are clear enough. So far, this is just a promissory note, but there is no clear argument based on Nagel's example that a precisely and thoroughly worked-out probabilistic semantics cannot be given.[10]

10. However unsympathetic Nagel's particular objections may be, they do point to a very general and vexed issue. Ordinarily, we understand ourselves to be able to assign probabilities to claims because we know what those claims mean; Reichenbach's semantics requires that we can calculate probabilities for claims and, on that basis, say what they mean. That seems quite unintuitive. It becomes more intuitive as a translation procedure: Assume I already speak an interpreted language and I am giving a semantics for a different language, L. If the speakers of L assign a probability to one of their sentences in every epistemic situation that is identical to the probability I assign to one of mine, then I should assign my sentence as the meaning in my language of their sentence in L.

In any case, regardless of the success of a probabilistic semantics, there can be no doubt that the centrality of probability for epistemology and the specific claims about the interpretation of probability that Reichenbach makes in the book were an important impetus in turning English-language philosophy toward technical work in the foundations of probability. Reichenbach's univocality thesis (there is one concept of probability that can cover both logical and physical probability) and his frequency interpretation of probability were and remain controversial claims. Whether philosophers agree with Reichenbach or not on those claims, part of Reichenbach's overall philosophical point of view succeeded when those issues are taken up as central issues of a newly technical epistemology.

The second main theme of the book has already been mentioned, the argument for realism and rejection of positivism. Reichenbach's defense of realism and rejection of a phenomenalist positivism takes place in the book within the context of chapter 2's very general question of the "existence of the external world." Now, many of the logical empiricists thought that they had shown not how to solve the problem of the external world but rather in precisely what sense there was no problem of the external world to be solved. Importantly for the early reception of *Experience and Prediction*, a number of American pragmatists thought that pragmatism had shown the same thing. Thus, one theme of the early reviews of the book was to figure out why Reichenbach thought there was a problem to be solved rather than to provide a detailed critique of his solution.

Part of the perceived difficulty was internal to Reichenbach's text. The setting of the problem of the existence of the external world—Do we have any reason to believe that anything exists beyond our impressions?—in chapter 2 of the book seems to rely on epistemological claims about the role of impressions in knowledge that are denied in chapter 3. If one is, as Reichenbach reveals himself to be in chapters 3 and 4, a physicalist regarding impressions and a fallibilist regarding sensory knowledge, the very problem of the external world is not, according to several of Reichenbach's reviewers, a problem at all. This was a judgment rendered by philosophers as disparate in their own projects as

William Barrett and Ernest Nagel. Barrett, writing in the *Journal of Philosophy* in 1939, claimed:

> Now one would not expect it to be likely that acceptance of the physicalistic basis for the language of science would lead us to raise the problem of an external world, which belongs to that quite distinct phase of positivism, in which it attempted to reduce knowledge to an absolutely certain basis in impressions and atomic facts. Certainly, if Reichenbach had already considered the subjects treated of in Chapter III of his book, *An Inquiry Concerning Sensations* (where he shows impressions to be abstracta from physical bodies), he could not have engaged in Chapter II in the long exposition of the problem of an external world. (Barrett 1939, 349)

At almost the same moment, in *Philosophy of Science*, Nagel wrote:

> Professor Reichenbach's discussion of the problem of the external world is a *tour de force*, almost by his own confession, and serves merely to exhibit the power of his theory of probability. For the discussion was generated by the assumption that impression propositions are absolutely certain. But Professor Reichenbach denies that they have this indubitable character, and considers in detail their more dubious epistemic status. (Nagel 1939, 240)

While the overall structure of Reichenbach's account of the external world problem remained something of a puzzle to many of his early readers (I will have a bit more to say about this in the next section, by way of indicating open topics of research for historical scholarship), Reichenbach's book had a resurgence of influence in the 1970s and into the contemporary literature thanks to the overall structure of his realist argument. Whatever the merits are of establishing the existence of an external world, the idea that one could combine probabilistic arguments and causal reasoning in defense of *scientific* realism inspired the work of Reichenbach's great student and advocate, Wesley Salmon (1979, 1994). That is to say, Salmon disentangled the structure of Reichenbach's realist argument from the epistemological setting of *Experience and Prediction* and used it not to overcome idealism and positivism so much as to oppose instrumentalist views of scientific theory. Within the arena of scientific realism, Salmon's advance of Reichenbach's general argument forms a species of "miracle argument" for scientific realism. On these views, the systematic regularities that the world presents to us in our experience are not equally well accounted for by realist arguments for theoretical entities and instrumentalist alternatives. Instrumentalism, rather, stands accused simply of asserting that there are such regularities and leaving them entirely unexplained (and, thus, miraculous).[11]

The third theme in the book that has generated much discussion is Reichenbach's pragmatic justification of induction, offered in the final chapter. The 1930s was a confident era in epistemology, and thus, solutions to Hume's problem of induction were not rare in those days. Reichenbach's solution may be the second most famous solution offered in the 1930s, outstripped only by Karl Popper's falsificationist solution. The very idea that Reichenbach's solution and Popper's solution are solutions to the same problem indicates one problem with "solving" the problem of induction—it is highly unclear how much of Hume's setting of the problem one must accept to be seen as solving rather than rejecting the problem. Popper's solution, for example, is a rather Pickwickian one: Popper agrees with Hume that if there were inductive inferences, we would not be justified in believing their conclusions. Popper then goes on to argue that this is not really a problem since there are no inductive inferences anyway—scientific generalizations are not conclusions of inferences from particulars but bold conjectures meant to cover individual cases and be tested against them.

Reichenbach's pragmatic justification of induction grants much more to Hume than does Popper's solution. Having granted that inductive inferences exist and are importantly employed in coming to general theories in science, Reichenbach proceeds (in section 38) to grant Hume two more claims about induction:

1. We have no logical demonstration for the validity of inductive inference.

11. An extremely valuable contribution to the literature on empiricism and scientific realism that helps explain both "miracle arguments" and the structure of Reichenbach's reasoning is Sober (1994).

2. There is no demonstration a posteriori for the inductive inference; any such demonstration would presuppose the very principle which it is to demonstrate.

These claims being granted, we can see what a conundrum the problem of induction is: We are to justify our use of inductive inference. Justification requires an argument for the reliability of inductive inference. Such an argument, if it is to be rationally persuasive, would, one would think, have to be deductively valid or inductively strong. But, we have just granted that there are no such arguments: deduction does not suffice to ground the inductive principle, and use of induction would beg the question.

In order to understand the general thrust of Reichenbach's reasoning and why he thinks a justification of induction is possible despite the argument just given—which he grants—it is useful to back up and consider a specific case of an inductive inference and see how he leads us through certain steps. Suppose, then, to take a familiar case, we have observed 100 black ravens and no nonblack ravens and we employ induction to conclude "all ravens are black." Reichenbach's first step is diagnostic and cautionary. He argues that when asked to justify an inference such as this one, philosophers are tempted to seek an ontological truth about the way the world is. Thus, many philosophers try to justify such inferential habits by arguing that we live in a world in which things divide into kinds with essential properties, or in which there are true laws of nature, or what have you. Reichenbach argues that, of course, if you knew such things about the world, induction would be justified, but since induction is our only reason to believe such things, the whole ontological grounding of inductive inference is a nonstarter.

Having discerned what we ought not be doing, Reichenbach then seeks to clarify the proper and general nature of the inference itself. This places the inference, according to Reichenbach, into its proper probabilistic setting. In essence, he argues that the inference about ravens above is this: The observed relative frequency of blackness among ravens for an initial sequence of 100 ravens is 1.00; thus, the relative frequency of blackness among ravens is 1.00. The fact that the probability involved equals 1 is not interesting; this setting of the problem indicates that what is at issue is the reliability of the inference of a probability judgment from an observed sequence to an extension of that sequence. The principle of induction for Reichenbach just says that the relative frequency for a sequence, no matter how long it is continued, is in a narrow range centered on the observed frequency (section 38).

Having discovered this probabilistic account of the principle of induction, Reichenbach does not set out to prove that it is true. Rather, he makes his "epistemological" and "pragmatic" turn and argues roughly as follows: Suppose we agree that the point of belief formation and knowledge is prediction. Suppose we agree that we do not know whether we live in a world in which there are true probabilistic generalizations and, thus, reliable predictions of the future based on observed frequencies. We can still argue as follows: Regardless of whether we live in a world with true probabilistic generalizations, if we employ the principle of induction, we will arrive at true generalizations *if there are any*. If we could show that the inductive principle is true, that would be sufficient to show that our inductive practices are justified, but, argues Reichenbach, we do not need to show that the inductive principle is true. Rather, once we agree on what the point of knowledge of the world is, we need only show that regardless of the way the world is, if we act according to the inductive principle we will get to the true generalizations if there are any. This is the sense in which the solution is "pragmatic": we'll have epistemic success, if there is any success to be had, if we act according to the principle of induction.

Beginning with the earliest reviews of the book, much ink has been spilled over whether Reichenbach's general strategy would provide a "solution" to the problem of induction even if successfully carried through and over whether he successfully carried it through. We cannot review all the issues involved in this literature.[12] One issue seems salient and general enough to pause over, however.

The issue involves the nature of the undischarged hypothetical: Reichenbach cannot establish that the world is predictable at all and establishes at most that if it is predictable, the principle of induction will lead

12. A richer introduction to the range of issues involved can be found in Salmon (1991).

to the true predictions. In an early set of objections to Reichenbach, Everett J. Nelson pressed hard on the sense in which we are to understand the antecedent of the conditional, given that Reichenbach admits that it cannot be established. Nelson, in essence, wanted to know why, if we cannot know whether we will ever succeed in prediction, we ought to act in conformity with the principle of induction. If it is, as it seems to be in Reichenbach's argument, simply because we want to be able to predict, then why should our mere desire for something make it reasonable to act as if that desire could be fulfilled, given that we have agreed we cannot know that?

The issue comes down to one of the conditions of rationality of action that aims at a goal that one cannot know is attainable. Reichenbach, at the time, enjoyed illustrating the logic of his justification of induction through use of an analogy involving shipwrecked mariners: Suppose a group of people were shipwrecked on an island that had no other sources of food, but they had some nets they could use for fishing (1938, 129). Reichenbach took it to be fairly obvious that even if they could not assign a nonzero probability to there being fish in the areas they could fish in, their desire and need for food commended to them that they go fishing. Fishing is the only thing they can do that might be successful; they will certainly die if they do not fish.

Reichenbach seems right about that, but there is a further issue. The reason we find it obvious that they ought to fish is that we know that there are fish in the sea that we can catch by fishing. Thus, while our poor people may be ignorant of any numerical probability that fishing will succeed in their area, their and our only sense that fishing *might* be successful depends, as Reichenbach allows (1938, 129), on our shared inductive knowledge that casting the nets in the ocean might succeed while tossing the nets in the air or praying to the gods or sitting down and staring into space will not. The question is really whether their specific ignorance regarding the local fish is sufficiently like the global ignorance of a predictable structure of the world for us to be able to use one to model the other. Here, my intuitions do not so much agree with or differ from Reichenbach's or Everett's as simply run out: in a world in which I could genuinely not know whether prediction would ever work,

I am not even certain I could employ a notion of prediction or action at all.[13]

There is one further element of the reaction to Reichenbach's solution to the problem of induction that bears mention. Nelson and, following him, Isabel Creed (1940) found a similarity between Reichenbach's argument structure and William James's argument in "The Will to Believe" (James [1896] 1956). Nelson brings up the issue in a discussion of the philosophical status of the principle: "even though I cannot know that the world is predictable, I should act as if the inductive principle is true." After noting that it cannot plausibly be a logical principle, Nelson asks:

> What is it, what kind of principle is it, and what is the evidence for it? (1) Empirical? No, for that would presuppose the solution to the inductive problem. (2) Metaphysical? No, for Professor Reichenbach is a metaphysical sceptic. (3) Syntactical, or linguistic? Hardly, because it is not a matter of tautological transformation. (4) Ethical? Has Professor Reichenbach resurrected James' "Will to Believe"? Or (5) what? (Nelson 1938, 360)

Clearly, the most apt answer is (4): although the principle is not narrowly ethical, it is practical in the sense of Kant—it is a hypothetical imperative. It posits an end and commends action sufficient for its attainment, if it can be attained at all. Moreover, the principle of induction is reliably sufficient in that case and, thus, fulfils the conditions of knowledge and not merely prophecy. The connection to James is quite evident: even though the setting in which the principle is enunciated

13. There is another disanalogy that Elliott Sober and Don Howard have stressed in comments on an earlier draft: Our fishermen probably have, in Reichenbach's scenario, only one means of catching fish. But, in the absence of an argument that the principle of induction is the only principle that will work, Reichenbach has not yet offered any account of why to prefer the use of it to the use of other working methods. There is a rich literature on other principles meeting the requirements of Reichenbach's vindication of induction. It would be in line with Reichenbach's pragmatism for him to view rival principles here as conventional or bifurcational options, subject to free choice. Regardless of such issues, that Reichenbach's work on the vindication of induction can still inspire technical work of the highest caliber is evidenced in, for example, Sober (1996) and Niiniluoto (1994).

could scarcely be more different from James's religious setting, Reichenbach argues, as does James, that unless one acts as if a certain claim were true independently of any evidence for its truth, a whole class of potential knowledge claims remain strictly outside one's epistemic horizons. This is not so much a will to believe as a will to act in a way that is necessary for empirical knowledge to be possible at all.[14]

This is to say that the justification of induction underscores the voluntarism that is so central to Reichenbach's account of knowledge. The voluntarism goes all the way down, so to speak. For the connection between knowledge and prediction is itself a choice and not a thesis of epistemology.[15] If one chooses to adopt a certain epistemic goal, one must act on the basis of a principle that one cannot know to be true. If one finds such a view incoherent, one has stopped agreeing with Reichenbach right back in section 1 of the book.[16]

The final theme is perhaps the most enduring aspect of the work. When called upon to explain what epistemological analysis is early in the work (section 1), Reichenbach introduced into philosophical parlance the distinction between the context of discovery and the context of justification. The epistemologist grants an arational context in which scientific theories are first thought up and worked out—the context of discovery. This context, precisely because it concerns a form of arational creativity, is not at issue in epistemology. It can be studied from an empirical point of view by psychologists, sociologists, and biographers; but it is not the context in which the rationality of science is displayed, and epistemologists have nothing to say about it. Epistemologists, who are

14. Pending an argument that the principle of induction is the only reliable principle, the necessity here claimed must be seen as internal, not external. You do not have to choose the inductive principle to have reliable predictive knowledge of the world, if other principles lead to that situation also, but to have reliable predictive knowledge of the world you must choose some such principle and act according to it.

15. That is, the tight connection of knowledge and prediction is a choice made by Reichenbach; it expresses his fundamental choice of aim of knowledge. It has no further justification.

16. In this context, it is interesting to note that Bas van Fraassen's (2002) recent advocacy of a voluntarist empiricism rests upon a history of empiricism in which James and Reichenbach find pride of place.

interested in the rationality of science, must look to the context of justification. This context appears in the efforts of scientists to communicate and justify their creatively discovered theories before their peers and finds its proximate form in the published scientific essay. Philosophers, on Reichenbach's view, possess a tool in formal logic that allows them to make the context of justification more rationally transparent, by casting scientific theories and arguments into more logically regimented language. Thus, logic is a tool by which the context of justification is rendered fully transparent, eventuating in the full rational reconstruction of scientific knowledge.

Reichenbach was not the first philosopher to distinguish the rational structure of knowledge from the psychosocial processes by which it is generated, but he was the first to express this distinction explicitly in terms of the distinction between the context of discovery and the context of justification. This distinction became the stock-in-trade of logical empiricists—it became the standard way in which they explained what they were doing as philosophers of science, even as it allowed them to agree that scientists had to have a sort of unbridled creativity in their theorizing. The context of justification, rendered fully explicit by the logical languages offered by technical philosophy of science, was where that creativity was placed under empirical control and rational scrutiny.

Having become standard doctrine under the logical empiricists, the distinction between the contexts of discovery and justification was the scene of much contestation in the 1950s and 1960s when philosophy of science opened itself up to new perspectives. In the 1950s, Norwood Russell Hanson tried to recover discovery as a topic in the philosophy of science in his book *Patterns of Discovery*. The distinction is one of the doctrines associated with logical empiricism that Kuhn argued against in his seminal "revolution-making" text, *The Structure of Scientific Revolutions* (Kuhn [1962] 1996). Given the centrality of issues relating to discovery in Kuhn's project—the difference between discovery in normal science and discovery in science in crisis, and the prevalence of multiple discovery—and given his objections to the account of science in scientific textbooks and "the philosophical works modeled on them" (Kuhn 1996, 136), one can easily see why Kuhn would not find anything analytically useful in insisting upon a distinction between the contexts of discovery and justification. He draws this out more explicitly in subsequent work, writing, for example, that "considerations relevant to the context of discovery are then relevant to justification as well; scientists who share the concerns and sensibilities of the individual who discovers a new theory are *ipso facto* likely to appear disproportionately frequently among that theory's first supporters" (Kuhn 1977, 328). In the late 1960s, Imre Lakatos ([1968] 1978) also made discovery a proper topic for a philosophy of science, by arguing that philosophy of science was the methodology of scientific research programs and by putting discovery practices into such programs under the name "positive heuristic."[17]

Experience and Prediction: New Interpretative Tasks

The themes just rehearsed are important for understanding the reception of Reichenbach's work. They retain their currency as issues in epistemology and philosophy of science and indicate how Reichenbach's work of 1938 might still provide resources for epistemology and philosophy of science.[18] Within the past twenty years or so, however, there has arisen another, more directly historical, project that seeks to understand the historical significance of logical empiricism for the development of philosophy of science and of analytic philosophy more generally. I shall close this introduction by rehearsing three issues for such historical understanding that are raised by *Experience and Prediction*, a book in which Reichenbach took pains to explain his version of logical empiricism to an Anglophone audience.

17. In accord with the odd enfoldings of philosophy, of course, Lakatos ends up "rationally reconstructing" the processes of discovery in his history of science. Thus, his relationship to Reichenbach is complicated, but we cannot go into the complications here.

18. I have argued that Reichenbach's characterization of the context of justification has more nuance and interest than most uses after Reichenbach (Richardson 2000) and therefore that Reichenbach's distinction is still a resource in helping philosophers of science understand their own projects—a resource that most arguments against the distinction do not really touch, given Reichenbach's complex setting for the distinction.

First, it is clear that most of the American readers of *Experience and Prediction* found it hard to square the projects of the three middle chapters of the book. The project of chapter 2—the solution to the external world problem—was understood to be an intervention in a hoary old problem, a problem that no longer moved many of the book's American readers. The project of chapter 3—an argument for the physical status of impressions and against their use as a certain foundation of knowledge—seemed to many readers to be more in line with current epistemological trends but also to be hard to square with the project of chapter 2. Chapter 4's project—the projective construction of the world on the basis of immediate things—received, except for some extensive but deeply puzzled remarks by Nagel (1939), almost no attention at all. The interpretative puzzle all this suggests is, of course, how to make sense of the epistemology on offer in these three chapters.

I suggest that we now have resources for thinking about that issue that were not widely available to the early American readers of the book. I argue, in particular, that the project of chapter 4 formed the key feature of Reichenbach's 1938 epistemology but that his American readers missed the point because chapter 4 is written as an elaboration upon and rejection of key features of Rudolf Carnap's 1928 *Der logische Aufbau der Welt* (*The Logical Structure of the World*) (Carnap [1928] 1967). In 1938 Carnap's book had not yet been translated into English, and Carnap himself, though ensconced in Chicago, was no longer pursuing the project of the *Aufbau*. Moreover, where the *Aufbau* was being read (by Quine and Goodman at Harvard, for example), it was given an interpretation (as an empiricist solution to the problem of the external world!) that would make it hard to see the deep interconnections between that work and Reichenbach's chapter 4.[19]

Reichenbach's main project in chapter 4 is the construction, on the basis of the immediately given objects of experience, of the distinction between the psychological and the physical realms. He follows Carnap

19. Carnap is referred to only once in chapter 2, and there only for the idea that philosophical differences involve a choice of different languages. Reichenbach, however, refers repeatedly to Carnap both in the early framing of the tasks of epistemology and in the chapter 4 project.

and others in denying that this distinction has already been drawn in choosing experience as the basis of knowledge—what is given in experience is not psychological or mental. Indeed, the whole chapter is a long exercise in constructing, within the structure of knowledge, the objective/subjective distinction. This is a project that animates much German epistemology, from Ernst Cassirer's neo-Kantianism to Edmund Husserl's transcendental phenomenology, but Reichenbach's general setting of the problem is clearly Carnapian.

The second issue follows from this one. The mental/physical distinction of chapter 4 is one place where a subjective/objective distinction is invoked or constructed in the book. Another place is very early in the meta-epistemological discussions of section 1, where Reichenbach assimilates the truth-directed versus volitional distinction to the objective/subjective distinction. He expresses the importance of his notion of entailed decisions as follows: "The concept of entailed decisions, therefore, may be regarded as a dam erected against extreme conventionalism; it allows us to separate the arbitrary part of the system of knowledge from its substantial content, to distinguish the subjective and the objective part of science" (section 1). Here, the subjective is identified with the arbitrary in the strict sense, with that which is due to the will of the epistemic agent or community. This sort of subjectivity cannot be eliminated from the structure of knowledge; Reichenbach's conventionalism, after all, demands that any structure of knowledge begin with decisions as to the meanings of terms and the goals of knowledge.[20]

Reichenbach's remarks on this sort of volitional subjectivity form an interesting episode in the history of early-twentieth-century accounts of the a priori element in knowledge. Reichenbach's earliest accounts of physical methodology were couched in a robustly Kantian idiom (Reichenbach [1920] 1965); decisions and definitions were necessary for knowledge and thus played the role of the Kantian a priori. Of course,

20. The philosopher of science can domesticate this subjectivity by engaging in "the advisory task" of epistemology, in which the subjective decisions are first scrutinized in second intention regarding their consequences and then offered as options to the epistemic community. This creates a generalized or collective will but does not eliminate the need for subjective decisions in the sense of "subjective" Reichenbach is here using.

unlike Kant, Reichenbach relativized the a priori: there was no single categorial framework necessary for objectivity to be possible. Under pressure from Moritz Schlick, Reichenbach, by the mid-1920s, ceased speaking in Kantian language. But, his general conventionalism of 1938 still bears the marks of the Kantian idea: the volitional decisions must be made before any system of objective knowledge is possible, and different choices will yield different structures. This is still a form of "relativized constitutive a priori" now presented in a robustly "pragmatic" idiom.[21]

It is interesting to compare Reichenbach's account here with that of C. I. Lewis, the great defender of a "pragmatic a priori" in America in the 1920s and 1930s. Lewis was less chary of using the term "a priori" than was Reichenbach, and Lewis's pragmatic a priori consists exactly in the volitional act in which a system of categories is chosen to apply to the world. Lewis posits a pure logical a priori, given in systems of concepts and imposed upon a pure world of experience. The pragmatic a priori involves two components: the need for some categorial framework to be imposed upon the given in order for the given to become the material of knowledge; and the fact that the choice of categorial framework is unconstrained by the given and a matter of free choice. Thus, regarding the pure systems of concepts and the pure world of fact, Lewis states:

> It is between these two, in the choice of the conceptual system for application and in the assigning of sensuous denotation to the abstract concept, that there is a pragmatic element in truth and knowledge. In this middle ground of trial and error, of expanding experience and the continual shift and modification of conception in our effort to cope with it, the drama of human interpretation and the control of nature is forever played out. (Lewis [1929] 1956, 272)

Lewis's version of a "pragmatic a priori" is in some ways less radical than Reichenbach's: Lewis is at pains to say that the world of experience, the given facts, are not altered by different conceptual frameworks.

21. On Reichenbach's importance in the story of a relativized Kantian a priori, see Friedman (1999, 2001).

Reichenbach's view, especially given his convention/bifurcation distinction, will have to be more subtle—Reichenbach seems to agree that conventional alternatives capture "the same facts," but this cannot be true of volitional bifurcations, which seem to induce an internal notion of factuality. Nonetheless, there is a family resemblance here that goes well beyond the views of these two thinkers to include conceptions of the conventional and the a priori developed in the 1920s and 1930s by philosophers as disparate as Rudolf Carnap, Charles Morris, and John Dewey.[22] Particularly striking in the case of Reichenbach and Lewis is their explicit talk of will or volition or purpose in presenting the role of the a priori and their commensurate reliance on a sort of subjectivity both pragmatic and transcendental. This element can also be found in Carnap, in his discussion of "external questions" and the pragmatic advice to accept or reject a certain language for science. The role of the will as the locus of the a priori has not been adequately theorized across the spectrum of early-twentieth-century scientific epistemologists.

Finally, one of the largest issues that has begun to occupy historians of twentieth-century philosophy is the reception of logical empiricism in America, particularly the relations between American pragmatism and logical empiricism.[23] *Experience and Prediction* is a central document for that issue. Reichenbach was not merely seeking American attention for the work by publishing it in English; he took pains to make reference to points of contact between his views and American pragmatism. Indeed, Reichenbach's list, given in the preface, of the philosophical movements that played a role in the origins of logical empiricism begins with "American pragmatists and behaviorists." We have already noted his "pragmatic solution" to the problem of induction (and its relations to James) and noted some elements common to his account of volition

22. I have dealt with the theories of the a priori in these thinkers more extensively in Richardson 2003a and 2003b.

23. This issue has been raised by Ronald N. Giere (1996) in the context of the history of philosophy of science in America and has been placed in a larger sociopolitical history of American philosophy recently by John McCumber (2001), Don Howard (2003), and George Reisch (2005). My remarks here are not by way of objection to the point of view of McCumber, Howard, and Reisch, which I find importantly correct, but by way of indicating how complicated the facts of the matter are.

in knowledge and accounts of the a priori found among pragmatists such as Lewis, Morris, and Dewey. Reichenbach begins chapter 3 by noting that his rejection of impressions as "observable facts" has much in common with arguments Dewey puts forth in *Experience and Nature* as well as with the behaviorism of E. C. Tolman.[24]

All these facts, together with Reichenbach's arrival in America almost simultaneously with the publication of his book, led several commentators to remark on the similarities and differences between Reichenbach's project and the American pragmatism of the late 1930s. The overall assessment was mixed. Writing for a learned public in *The Nation*, Sidney Hook opined that "the American reader will be impressed by the fact that [Reichenbach] seems to have developed views which are close to the probabilism of Charles Peirce" (1939, 40). But Hook was not entirely sure what to make of the overall project of the book:

> A considerable portion of the book is devoted to epistemological questions, including the grounds of belief in the existence of the external world, which pragmatic American philosophers with whom Reichenbach claims kinship have dismissed as meaningless. Professor Reichenbach's absorption in traditional epistemological problems, combined with his operationalism and his frequency theory of probability, creates some puzzles that will stir many American philosophers to criticism and dissent. But his work will undoubtedly have a fructifying influence. (Hook 1939, 41)

Ernest Nagel was equally ambivalent in his assessment of Reichenbach's work and the "affiliations of European philosophy of science and our home-grown pragmatism" (Nagel 1938, 272). Nagel ended his review with an image of Reichenbach as something of an oddity:

> Reichenbach is sympathetic to pragmatists, although he is not sure if he is one himself; certainly his resuscitation of epistemological problems buried long ago by American pragmatists make him a strange and unique specimen.

24. One can only conjecture as to whether Reichenbach's title was chosen in self-conscious connection with Dewey's *Experience and Nature*. Certainly, the substitution of "prediction" for "nature" is suggestive of the differences between Dewey's project and Reichenbach's.

> His book gives evidence that although pragmatists have something to learn from him, he too has something to learn from them. (Nagel 1938, 272)

While the American philosophers certainly had a mixed reaction to the work, one type of reaction was not to be seen. A story one hears is that logical empiricism brought a narrow-minded professional and technical philosophy to the United States and upset a more expansive and socially aware indigenous philosophy. It is very difficult to find any objections to the technical nature of Reichenbach's epistemology in the early reviews,[25] and indeed, as Hook's comments show, technicality can be seen as a point of contact between Reichenbach and the Peircean wing of pragmatism. The concerns expressed throughout the early responses were not about the technical aspects of the work but about the epistemological projects those technical aspects served. We have seen Nagel inveighing against those projects; here is Nagel's advice to Reichenbach and to the community of philosophers in America:

> Having now made his venture into epistemology, will not Professor Reichenbach return to the kind of analyses he was performing when he wrote his *Axiomatik der Raum-Zeit Lehre* and his *Philosophie der Raum-Zeit Lehre* [*sic*]? It is in such works that the strength and virtue of scientific philosophy lie. (Nagel 1939, 253)

In 1939, Reichenbach, now safely on American soil at UCLA, published an essay on John Dewey's theory of science. He ended the essay with this call to scientific organization of philosophical work:

> The early period of empiricism in which an all-round philosopher could dominate at the same time the fields of scientific method, of history of philosophy, of education and social philosophy, has passed. We enter into the second phase in which highly technical investigations form the indispensable instruments of research, splitting the philosophical campus into spe-

25. The one exception I have seen was in the review by Floyd N. House in *American Sociological Review*. The philosophers did not share House's worry about the project's "technical character" or Reichenbach's "abstruse terminology" (House 1939, 580).

> cialists of its various branches. We should not regret this unavoidable specialization which repeats on philosophic grounds a phenomenon well known from all other fields of scientific inquiry. (Reichenbach 1939, 192)

Reichenbach may have understood himself to be admonishing his new American colleagues to reconfigure the research community in philosophy; but none of the American philosophical reviewers of *Experience and Prediction* seemed threatened by such an idea, and, indeed, Nagel had argued already that Reichenbach's general epistemology ran counter to the spirit of a specialized and divided research community.[26]

Notwithstanding Nagel's sensibility, one leading task of the new history of logical empiricism is to connect the technical projects characteristic of logical empiricist philosophy of science with more general philosophical visions that give those projects form and content. Recent literature has addressed a variety of such logical empiricist philosophical visions, including those found in Carnap (Richardson 1998) and Neurath (Uebel 1991); Reichenbach's explicit attempt in *Experience and Prediction* to provide the more general epistemological framework for his technical work makes the present book invaluable in reflecting on the philosophy of the twentieth century. No work better illuminates Reichenbach's distinctive philosophical point of view. This is reason enough to commend it to the attention of today's philosophical community.

References Cited

Barrett, William. 1939. "On the Existence of an External World." *Journal of Philosophy* 36: 346–54.

26. We ought to be clear here. First, Reichenbach certainly meant to disrupt what he understood to be the division of philosophical labor in the United States; this is shown in the above quotation about Dewey. Second, the absence of a clear expression in the early American reviews of Reichenbach that his account of philosophy did disrupt their community structure is not in itself evidence that Reichenbach views and his activities did not have that consequence. Proper investigation of this historical claim, however, would require a much deeper institutional history of twentieth-century American philosophy than anyone has yet written.

Bisbee, Eleanor. 1938. "A World of Probability." [Review of *Experience and Prediction.*] *Philosophy of Science* 5: 360–66.

Carnap, Rudolf. [1928] 1967. *The Logical Structure of the World.* Berkeley and Los Angeles: University of California Press.

———. [1934] 1937. *Logical Syntax of Language.* London: Routledge and Kegan Paul.

———. 1936/1937. "Testability and Meaning." *Philosophy of Science* 3: 419–71; 4: 1–40.

Creed, Isabel. 1940. "The Justification of the Habit of Induction." *Journal of Philosophy* 37: 85–97.

Dennes, William R. 1939. [Review of *Experience and Prediction.*] *Philosophical Review* 48: 536–38.

Feigl, Herbert, and Alfred E. Blumberg. 1931. "Logical Positivism." *Journal of Philosophy* 28: 281–96.

Friedman, Michael. 1999. *Reconsidering Logical Positivism.* Cambridge: Cambridge University Press.

———. 2001. *The Dynamics of Reason.* Stanford: CSLI Publications.

Giere, Ronald N. 1996. "From *wissenschaftliche Philosophie* to Philosophy of Science." In R. N. Giere and A. W. Richardson, eds., *Origins of Logical Empiricism*, 335–54. Minneapolis: University of Minnesota Press.

Hanson, Norwood Russell. 1958. *Patterns of Discovery.* Cambridge: Cambridge University Press.

Hook, Sidney. 1939. "Logical Empiricism." [Review of *Experience and Prediction.*] *The Nation* 148: 40–41.

House, Floyd N. [Review of *Experience and Prediction.*] *American Journal of Sociology* 44: 579–80.

Howard, Don. 1994. "Einstein, Kant, and the Origins of Logical Empiricism." In W. Salmon and G. Wolters, eds., *Logic, Language, and the Structure of Scientific Theories*, 45–105. Pittsburgh and Konstanz: University of Pittsburgh Press/ Universitätsverlag Konstanz.

———. 2003. "Two Left Turns Make a Right: On the Curious Political Career of North American Philosophy of Science at Midcentury." In G. L. Hardcastle and A.W. Richardson, eds., *Logical Empiricism in North America*, 25–93. Minneapolis: University of Minnesota Press.

James, William. [1896] 1956. *The Will to Believe and Human Immortality.* New York: Dover.

Kuhn, Thomas S. [1962] 1996. *The Structure of Scientific Revolutions.* Chicago: University of Chicago Press.

———. 1977. "Objectivity, Value Judgment, and Theory Choice." In *The Essential Tension*, 320–39. Chicago: University of Chicago Press.

Lakatos, Imre. [1968] 1978. "Falsification and the Methodology of Scientific Research Programmes." In John Worrall and Gregory Curry, eds., *The Methodology of Scientific Research Programmes*, 8–101. Cambridge: Cambridge University Press.

Lewis, C. I. [1929] 1956. *Mind and the World Order.* New York: Dover.

McCumber, John. 2001. *Time in the Ditch: American Philosophy and the McCarthy Era.* Evanston: Northwestern University Press.

Nagel, Ernest. 1938. [Review of *Experience and Prediction.*] *Journal of Philosophy* 35: 270–72.

———. 1939. "Probability and the Theory of Knowledge." *Philosophy of Science* 6: 212–53.

Nelson, Everett J. 1938. "Comments and Criticisms." [Reply to Reichenbach.] *Journal of Philosophy* 35: 355–60.

Niiniluoto, Ilkka. 1994. "Descriptive and Inductive Simplicity." In W. Salmon and G. Wolters, eds., *Logic, Language, and the Structure of Scientific Theories*, 147–70. Pittsburgh: University of Pittsburgh Press.

Poincaré, Henri. [1902] 1946. *Science and Hypothesis.* New York: Dover.

Reichenbach, Hans. 1916/1917. "Der Begriff der Wahrscheinlichkeit für die mathematische Darstellung der Wirklichkeit." *Zeitschrift für Philosophie und philosophische Kritik* 161: 210–39; 162: 9–112, 223–53.

———. [1920] 1965. *The Theory of Relativity and A Priori Knowledge.* Berkeley and Los Angeles: University of California Press.

———. [1928] 1958. *Philosophy of Space and Time.* New York: Dover.

———. 1932. *Atom and Cosmos.* London: George Braziller.

———. 1935. *Wahrscheinlichkeitslehre: Eine Untersuchung über die logischen und mathematischen Grundlagen der Wahrscheinlichkeitsrechnung.* Leyden: Sijthoff.

———. 1936. "Logical Empiricism in Germany and the Present State of Its Problems." *Journal of Philosophy* 33: 141–60.

———. 1938. "Comments and Criticism." [Reply to Nelson.] *Journal of Philosophy* 35: 127–30.

———. 1939. "Dewey's Theory of Science." In P. A. Schilpp, ed., *The Philosophy of John Dewey*, 159–92. Evanston: Northwestern University Press.

———. 1951. *The Rise of Scientific Philosophy.* Berkeley and Los Angeles: University of California Press.

———. 1978. *Selected Writings: 1909–1953.* Dordrecht: Reidel, 1978.

Reisch, George A. 2005. *How the Cold War Transformed Philosophy of Science: To the Icy Slopes of Logic.* New York: Cambridge University Press.

Richardson, Alan W. 1998. *Carnap's Construction of the World.* Cambridge: Cambridge University Press.

———. 2000. "Science as Will and Representation: Carnap, Reichenbach, and the Sociology of Science." *Philosophy of Science (Proceedings)* 67: S151–S162.

———. 2003a. "Logical Empiricism, American Pragmatism, and the Fate of Scientific Philosophy in North America." In G. L. Hardcastle and A. W. Richardson, eds., *Logical Empiricism in North America*, 1–24. Minneapolis: University of Minnesota Press.

———. 2003b. "The Geometry of Knowledge: Becker, Carnap, and Lewis and the Formalization of Philosophy in the 1920s." *Studies in History and Philosophy of Science* 34: 165–82.

Ryckman, Thomas A. 1992. "P(oint) C(oincidence) Thinking: The Ironical Attachment of Logical Empiricism to General Relativity (and Some Lingering Consequences)." *Studies in History and Philosophy of Science* 23: 471–93.

———. Forthcoming. "Logical Empiricism and the Philosophy of Physics." In T. E. Uebel and A. W. Richardson, eds., *The Cambridge Companion to Logical Empiricism.* Cambridge: Cambridge University Press.

Salmon, Wesley. 1979. "The Philosophy of Hans Reichenbach." In W. Salmon, ed., *Hans Reichenbach: Logical Empiricist*, 1–84. Dordrecht: Reidel.

———. 1991. "Hans Reichenbach's Vindication of Induction." *Erkenntnis* 39: 99–122.

———. 1994. "Carnap, Hempel, and Reichenbach on Scientific Realism." In W. Salmon and G. Wolters, eds., *Logic, Language, and the Structure of Scientific Theories*, 237–54. Pittsburgh and Konstanz: University of Pittsburgh Press/ Universitätsverlag Konstanz.

Sklar, Lawrence. 1974. *Space, Time, and Space-Time.* Berkeley and Los Angeles: University of California Press.

Sober, Elliott. 1994. "Contrastive Empiricism." In his *From a Biological Point of View*, 114–39. Cambridge: Cambridge University Press.

———. 1996. "Parsimony and Predictive Equivalence." *Erkenntnis* 44: 167–97.

Stadler, Friedrich. 1997. *Studien zum Wiener Kreis.* Frankfurt: Suhrkamp.

Uebel, Thomas E. 1991. *Overcoming Logical Empiricism from Within.* Amsterdam: Rodopi.

van Fraassen, Bas C. 2002. *The Empirical Stance.* New Haven: Yale University Press.

Werkmeister, W. H. 1938. [Review of, among others, *Experience and Prediction.*] *Ethics* 48: 549–54.

Experience and Prediction

PREFACE

The ideas of this book have grown from the soil of a philosophic movement which, though confined to small groups, is spread over the whole world. American pragmatists and behaviorists, English logistic epistemologists, Austrian positivists, German representatives of the analysis of science, and Polish logisticians are the main groups to which is due the origin of that philosophic movement which we now call "logistic empiricism." The movement is no longer restricted to its first centers, and its representatives are to be found today in many other countries as well—in France, Italy, Spain, Turkey, Finland, Denmark, and elsewhere. Though there is no philosophic system which unites these groups, there is a common property of ideas, principles, criticisms, and working methods—all characterized by their common descent from a strict disavowal of the metaphor language of metaphysics and from a submission to the postulates of intellectual discipline. It is the intention of uniting both the empiricist conception of modern science and the formalistic conception of logic, such as expressed in logistic, which marks the working program of this philosophic movement.

Since this book is written with the same intentions, it may be asked how such a new attempt at a foundation of logistic empiricism can be justified. Many things indeed will be found in this book which have been said before by others, such as the physicalistic conception of language and the importance attributed to linguistic analysis, the connection of meaning and verifiability, and the behavioristic conception of psychology. This fact may in part be justified by the intention of giving a report of those results which may be considered today as a secured possession of the philosophic movement described; however, this is not the sole intention. If the present book enters once more into the discussion of these fundamental problems, it is because former investigations did not sufficiently take into account one concept which penetrates into all the logical relations constructed in these domains: that is, the concept of probability. It is the intention of this book to show the fundamental place which is occupied in the system of knowledge by this concept and to point out the consequences involved in a consideration of the probability character of knowledge.

The idea that knowledge is an approximative system which will never become "true" has been acknowledged by almost all writers of the empiricist group; but never have the logical consequences of this idea been sufficiently realized. The approximative character of science has been considered as a necessary evil, unavoidable for all practical knowledge, but not to be counted among the essential features of knowledge; the probability element in science was taken as a provisional feature, appearing in scientific investigation as long as it is on the path of discovery but disappearing in knowledge as a definitive system. Thus a fictive definitive system of knowledge was made the basis of epistemological inquiry, with the result that the schematized character of this basis was soon forgotten, and the fictive construction was identified with the actual system. It is one of the elementary laws of approximative procedure that the consequences drawn from a schematized conception do not hold outside the limits of the approximation; that in particular no consequences may be drawn from features belonging to the nature of the schematization only and not to the co-ordinated object. Mathematicians know that for many a purpose the number π may be sufficiently approximated by the value 22/7; to infer from this, however, that π is a rational number is by no means permissible. Many of the inferences of traditional epistemology and of positivism as well, I must confess, do not appear much better to me. It is particularly the domain of the verifiability conception of meaning and of questions connected with it, such as the problem of the existence of external things, which has been overrun with paralogisms of this type.

The conviction that the key to an understanding of scientific method is contained within the probability problem grew stronger and stronger with me in the face of such basic mistakes. This is the reason why, for a long time, I renounced a comprehensive report of my epistemological views, although my special investigations into different problems of epistemology demanded a construction of foundations different from those constructed by some of my philosophical friends. I concentrated my inquiry on the problem of probability which demanded at the same time a mathematical and a logical analysis. It is only after having traced out a logistic theory of probability, including a solution of the problem of induction, that I turn now to an application of these ideas to questions of a more general epistemological character. As my theory of probability has been published for some years, it was not necessary to present it with all mathematical details once more in the present book; the fifth chapter, however, gives an abbreviated report of this theory—a report which seemed necessary as the probability book has been published in German only.

It is this combination of the results of my investigations on probability with the ideas of an empiricist and logistic conception of knowledge which I here present as my contribution to the discussion of logistic empiricism. The growth of this movement seems to me sufficiently advanced to enter upon a level of higher approximation; and what I propose is that the form of this new phase should be a probabilistic empiricism. If the continuation suggested comes to contradict some ideas so far considered as established, particularly by positivist writers, the reader will bear in mind that this criticism is not offered with the intention of diminishing the historical merits of these philosophers. On the contrary, I am glad to have an occasion for expressing my indebtedness to many a writer whose opinions I cannot wholly share. I think, however, that the clarification of the foundations of our common conceptions is the most urgent task within our philosophic movement and that we should not recoil from frankly admitting the insufficiencies of former results—even if they still find defenders within our ranks.

The ideas of this book have been discussed in lectures and seminars at the University of Istanbul. I welcome the opportunity to express my warmest thanks to friends and students here in Istanbul for their active interest which formed a valuable stimulus in the clarification of my ideas, especially to my assistant, Miss Neyire Adil-Arda, without whose constant support I should have found it very much harder to formulate my views. For help in linguistic matters and reading of proofs I am grateful to Miss Sheila Anderson, of the English High School at Istanbul; to Professor Charles W. Morris, Mr. Lawrence K. Townsend, Jr., and Mr. Rudolph C. Waldschmidt, of the University of Chicago; to Mr. Max Black, of the Institute of Education of the University of London; and to Miss Eleanor Bisbee, of Robert College at Istanbul.

Hans Reichenbach

University of Istanbul
Turkey

CHAPTER I

MEANING

CHAPTER I

MEANING

§ 1. The three tasks of epistemology

Every theory of knowledge must start from knowledge as a given sociological fact. The system of knowledge as it has been built up by generations of thinkers, the methods of acquiring knowledge used in former times or used in our day, the aims of knowledge as they are expressed by the procedure of scientific inquiry, the language in which knowledge is expressed—all are given to us in the same way as any other sociological fact, such as social customs or religious habits or political institutions. The basis available for the philosopher does not differ from the basis of the sociologist or psychologist; this follows from the fact that, if knowledge were not incorporated in books and speeches and human actions, we never would know it. Knowledge, therefore, is a very concrete thing; and the examination into its properties means studying the features of a sociological phenomenon.

We shall call the first task of epistemology its *descriptive task*—the task of giving a description of knowledge as it really is. It follows, then, that epistemology in this respect forms a part of sociology. But it is only a special group of questions concerning the sociological phenomenon "knowledge" which constitutes the domain of epistemology. There are such questions as "What is the meaning of the concepts used in knowledge?" "What are the presuppositions contained in the method of science?" "How do we know whether a sentence is true, and do we know that at all?" and many others; and although, indeed, these questions concern the sociological phenomenon "science," they are of a very special type as compared with the form of questions occurring in general sociology.

What makes this difference? It is usually said that this is a difference of internal and external relations between those human utterances the whole of which is called "knowledge." Internal relations are such as belong to the content of knowledge, which must be realized if we want to understand knowledge, whereas external relations combine knowledge with utterances of another kind which do not concern the content of knowledge. Epistemology, then, is interested in internal relations only, whereas sociology, though it may partly consider internal relations, always blends them with external relations in which this science is also interested. A sociologist, for instance, might report that astronomers construct huge observatories containing telescopes in order to watch the stars, and in such a way the internal relation between telescopes and stars enters into a sociological description. The report on contemporary astronomy begun in the preceding sentence might be continued by the statement that astronomers are frequently musical men, or that they belong in general to the bourgeois class of society; if these relations do not interest epistemology, it is because they do not enter into the content of science—they are what we call external relations.

Although this distinction does not furnish a sharp line of demarcation, we may use it for a first indication of the design of our investigations. We may then say the descriptive task of epistemology concerns the internal structure of knowledge and not the external features which appear to an observer who takes no notice of its content.

We must add now a second distinction which concerns psychology. The internal structure of knowledge is the system of connections as it is followed in thinking. From such a definition we might be tempted to infer that epistemology is the giving of a description of thinking processes; but that would be entirely erroneous. There is a great difference between the system of logical interconnections of thought and the actual way in which thinking processes are performed. The psychological operations of thinking are rather vague and fluctuating processes; they almost never keep to the ways prescribed by logic and may even skip whole groups of operations which would be needed for a complete exposition of the subject in question. That is valid for thinking in daily life, as well as for the mental procedure of a man of science, who is confronted by the task of finding logical interconnections between divergent ideas about newly observed facts; the scientific genius has never felt bound to the narrow steps and prescribed courses of logical reasoning. It would be, therefore, a vain attempt to construct a theory of knowledge which is at the same time logically complete and in strict correspondence with the psychological processes of thought.

The only way to escape this difficulty is to distinguish carefully the task of epistemology from that of psychology. Epistemology does not regard the processes of thinking in their actual occurrence; this task is entirely left to psychology. What epistemology intends is to construct thinking processes in a way in which they ought to occur if they are to be ranged in a consistent system; or to construct justifiable sets of operations which can be intercalated between the starting-point and the issue of thought-processes, replacing the real intermediate links. Epistemology thus considers a logical substitute rather than real processes. For this logical substitute the term *rational reconstruction* has been introduced;[1] it seems an appropriate phrase to indi-

[1] The term *rationale Nachkonstruktion* was used by Carnap in *Der logische Aufbau der Welt* (Berlin and Leipzig, 1928).

cate the task of epistemology in its specific difference from the task of psychology. Many false objections and misunderstandings of modern epistemology have their source in not separating these two tasks; it will, therefore, never be a permissible objection to an epistemological construction that actual thinking does not conform to it.

In spite of its being performed on a fictive construction, we must retain the notion of the descriptive task of epistemology. The construction to be given is not arbitrary; it is bound to actual thinking by the postulate of correspondence. It is even, in a certain sense, a better way of thinking than actual thinking. In being set before the rational reconstruction, we have the feeling that only now do we understand what we think; and we admit that the rational reconstruction expresses what we mean, properly speaking. It is a remarkable psychological fact that there is such an advance toward understanding one's own thoughts, the very fact which formed the basis of the mäeutic of Socrates and which has remained since that time the basis of philosophical method; its adequate scientific expression is the principle of rational reconstruction.

If a more convenient determination of this concept of rational reconstruction is wanted, we might say that it corresponds to the form in which thinking processes are communicated to other persons instead of the form in which they are subjectively performed. The way, for instance, in which a mathematician publishes a new demonstration, or a physicist his logical reasoning in the foundation of a new theory, would almost correspond to our concept of rational reconstruction; and the well-known difference between the thinker's way of finding this theorem and his way of presenting it before a public may illustrate the difference in question. I shall introduce the terms *context of discovery* and *context of justification* to mark this distinction. Then we have to say that epistemology is only occupied in constructing the context of justification. But even the way of presenting scientific theories is only an approximation to what we mean by the context of justification. Even in the written form scientific expositions do not always correspond to the exigencies of logic or suppress the traces of subjective motivation from which they started. If the presentation of the theory is subjected to an exact epistemological scrutiny, the verdict becomes still more unfavorable. For scientific language, being destined like the language of daily life for practical purposes, contains so many abbreviations and silently tolerated inexactitudes that a logician will never be fully content with the form of scientific publications. Our comparison, however, may at least indicate the way in which we want to have thinking replaced by justifiable operations; and it may also show that the rational reconstruction of knowledge belongs to the descriptive task of epistemology. It is bound to factual knowledge in the same way that the exposition of a theory is bound to the actual thoughts of its author.

In addition to its descriptive task, epistemology is concerned with another purpose which may be called its *critical task*. The system of knowledge is criticized; it is judged in respect of its validity and its reliability. This task is already partially performed in the rational reconstruction, for the fictive set of operations occurring here is chosen from the point of view of justifiability; we replace actual thinking by such operations as are justifiable, that is, as can be demonstrated as valid. But the tendency to remain in correspondence with actual thinking must be separated from the tendency to obtain valid thinking; and so we have to distinguish between the descriptive and the critical task. Both collaborate in the rational reconstruction. It may even happen that the description of knowledge leads to the result that certain chains of thoughts, or operations, cannot be justified; in other words, that even the rational reconstruction contains unjustifiable chains, or that it is not possible to intercalate a justifiable chain between the starting-point and the issue of actual thinking. This case shows that the descriptive task and the critical task are different; although description, as it is here meant, is not a copy of actual thinking but the construction of an equivalent, it is bound by the postulate of correspondence and may expose knowledge to criticism.

The critical task is what is frequently called *analysis of science;* and as the term "logic" expresses nothing else, at least if we take it in a sense corresponding to its use, we may speak here of the logic of science. The well-known problems of logic belong to this domain; the theory of the syllogism was built up to justify deductive thinking by reducing it to certain justifiable schemes of operation, and the modern theory of the tautological character of logical formulas is to be interpreted as a justification of deductive thinking as conceived in a more general form. The question of the synthetic a priori, which has played so important a role in the history of philosophy, also falls into this frame; and so does the problem of inductive reasoning, which has given rise to more than one "inquiry concerning human understanding." Analysis of science comprehends all the basic problems of traditional epistemology; it is, therefore, in the foreground of consideration when we speak of epistemology.

The inquiries of our book will belong, for the most part, to the same domain. Before entering upon them, however, we may mention a result of rather general character which has been furnished by previous investigations of this kind —a result concerning a distinction without which the process of scientific knowledge cannot be understood. Scientific method is not, in every step of its procedure, directed by the principle of validity; there are other steps which have the character of volitional decisions. It is this distinction which we must emphasize at the very beginning of epistemological investigations. That the idea of truth, or validity, has a directive influence in scientific thinking is obvious and has at all times been noticed by epistemologists. That there are certain elements of knowledge, however, which are not governed by the idea of truth, but which are due to volitional resolutions, and though highly influencing the makeup of the whole system of knowledge, do not touch its truth-character, is less known to philosophical investigators. The presentation of the volitional decisions contained in the system of knowledge constitutes, therefore, an integral part of the critical task of epistemology. To give an example of volitional decisions, we may point to the so-called *conventions*, e.g., the convention concerning the unit of length, the decimal system, etc. But not all conventions are so obvious, and it is sometimes a rather difficult problem to find out the points which mark conventions. The progress of epistemology has been frequently furthered by the discovery of the conventional character of certain elements taken, until that time, as having a truth-character; Helmholtz' discovery of the arbitrariness of the definition of spatial congruence, Einstein's discovery of the relativity of simultaneity, signify the recognition that what was deemed a statement is to be replaced by a decision. To find out all the points at which decisions are involved is one of the most important tasks of epistemology.

The conventions form a special class of decisions; they represent a choice between *equivalent* conceptions. The different systems of weights and measures constitute a good example of such an equivalence; they illustrate the fact that the decision in favor of a certain convention does not influence the content of knowledge. The examples chosen from the theory of space and time previously mentioned are likewise to be ranked among conventions. There are decisions of another character which do not lead to equivalent conceptions but to divergent systems; they may be called *volitional bifurcations*. Whereas a convention may be compared to a choice between different ways leading to the same place, the volitional bifurcation resembles a bifurcation of ways which will never meet again. There are some volitional bifurcations of an important character which stand at the very entrance of science: these are decisions concerning the aim of science. What is the purpose of scientific inquiry? This is, logically speaking, a question not of truth-character but of volitional decision, and the decision determined by the answer to this question belongs to the bifurcation type. If anyone tells us that he studies science for his pleasure and to fill his hours of leisure, we cannot raise the objection that this reasoning is "a false statement"—it is no statement at all but a decision, and everybody has the right to do what he wants. We may object that such a determination is opposed to the normal use of words and that what he calls the aim of science is generally called the aim of play—this would be a true statement. This statement belongs to the descriptive part of epistemology; we can show that in books and discourses the word "science" is always connected with "discovering truth," sometimes also with "foreseeing the future." But, logically speaking, this is a matter of volitional decision. It is obvious that this decision is not a convention because the two conceptions obtained by different postulates concerning the aims of science are not equivalent; it is a bifurcation. Or take a question as to the meaning of a certain concept—say, causality, or truth, or meaning itself. Logically speaking this is a question of a decision concerning the limitation of a concept, although, of course, the practice of science has already decided about this limitation in a rather precise way. In such a case, it must be carefully examined whether the decision in question is a convention or a bifurcation. The limitation of a concept may be of a conventional character, i.e., different limitations may lead to equivalent systems.

The character of being true or false belongs to statements only, not to decisions. We can, however, co-ordinate with a decision certain statements concerning it; and, above all, there are two types of statements which must be considered. The first one is a statement of the type we have already mentioned; it states which decision science uses in practice. It belongs to descriptive epistemology and is, therefore, of a sociological character. We may say that it states an *object fact*, i.e., a fact belonging to the sphere of the objects of knowledge,[2] a sociological fact being of this type. It is, of course, the same type of fact with which natural science deals. The second statement concerns the fact that, logically speaking, there is a decision and not a statement; this kind of fact may be called a *logical fact*. There is no contradiction in speaking here of a fact concerning a decision; although a decision is not a fact, its character of being a decision is a fact and may be expressed in a statement. That becomes obvious by the cognitional character of such a statement; the statement may be right or wrong, and in some cases the wrong statement has been maintained for centuries, whereas the right statement was discovered only recently. The given examples of Helm-

[2] The term "objective fact" taken in the original sense of the word "objective" would express the same point; but we avoid it, as the word "objective" suggests an opposition to "subjective," an opposition which we do not intend.

holtz' and Einstein's theories of space and time may illustrate this. But the kind of fact maintained here does not belong to the sphere of the objects of science, and so we call it a logical fact. It will be one of our tasks to analyze these logical facts and to determine their logical status; but for the present we shall use the term "logical fact" without further explanation.

The difference between statements and decisions marks a point at which the distinction between the descriptive task and the critical task of epistemology proves of utmost importance. Logical analysis shows us that within the system of science there are certain points regarding which no question as to truth can be raised, but where a decision is to be made; descriptive epistemology tells us what decision is actually in use. Many misunderstandings and false pretensions of epistemology have their origin here. We know the claims of Kantianism, and Neo-Kantianism, to maintain Euclidean geometry as the only possible basis of physics; modern epistemology showed that the problem as it is formulated in Kantianism is falsely constructed, as it involves a decision which Kant did not see. We know the controversies about the "meaning of meaning"; their passionate character is due to the conviction that there is an absolute meaning of meaning which we must discover, whereas the question can only be put with respect to the concept of meaning corresponding to the use of science, or presupposed in certain connections. But we do not want to anticipate the discussion of this problem, and our later treatment of it will contain a more detailed explanation of our distinction between statements and decisions.

The concept of decision leads to a third task with which we must charge epistemology. There are many places where the decisions of science cannot be determined precisely, the words or methods used being too vague; and there are others in which two or even more different decisions are in use, intermingling and interfering within the same context and confusing logical investigations. The concept of meaning may serve as an example; simpler examples occur in the theory of measurement. The concrete task of scientific investigation may put aside the exigencies of logical analysis; the man of science does not always regard the demands of the philosopher. It happens, therefore, that the decisions presupposed by positive science are not clarified. In such a case, it will be the task of epistemology to suggest a proposal concerning a decision; and we shall speak, therefore, of the *advisory task* of epistemology as its third task. This function of epistemology may turn out to be of great practical value; but it must be kept clearly in mind that what is to be given here is a proposal and not a determination of a truth-character. We may point out the advantages of our proposed decision, and we may use it in our own expositions of related subjects; but never can we demand agreement to our proposal in the sense that we can demand it for statements which we have proven to be true.

There is, however, a question regarding facts which is to be considered in connection with the proposal of a decision. The system of knowledge is interconnected in such a way that some decisions are bound together; one decision, then, involves another, and, though we are free in choosing the first one, we are no longer free with respect to those following. We shall call the group of decisions involved by one decision its *entailed decisions*. To give a simple example: the decision for the English system of measures leads to the impossibility of adding measure numbers according to the technical rules of the decimal system; so the renunciation of these rules would be an entailed decision. Or a more complicated example: the decision expressed in the acceptance of Euclidean geometry in physics may lead to the occurrence of strange forces, "universal forces," which distort all bodies to the same extent, and may lead to even greater inconveniences concerning the continuous character of causality.[3] The discovery of interconnections of this kind is an important task of epistemology, the relations between different decisions being frequently hidden by the complexity of the subject; it is only by adding the group of entailed decisions that a proposal respecting a new decision becomes complete.

The discovery of entailed decisions belongs to the critical task of epistemology, the relation between decisions being of the kind called a logical fact. We may therefore reduce the advisory task of epistemology to its critical task by using the following systematic procedure: we renounce making a proposal but instead construe a list of all possible decisions, each one accompanied by its entailed decisions. So we leave the choice to our reader after showing him all factual connections to which he is bound. It is a kind of logical signpost which we erect; for each path we give its direction together with all connected directions and leave the decision as to his route to the wanderer in the forest of knowledge. And perhaps the wanderer will be more thankful for such a signpost than he would be for suggestive advice directing him into a certain path. Within the frame of the modern philosophy of science there is a movement bearing the name of *conventionalism;* it tries to show that most of the epistemological questions contain no questions of truth-character but are to be settled by arbitrary decisions. This tendency, and above all, in its founder Poincaré, had historical merits, as it led philosophy to stress the volitional elements of the system of knowledge

[3] Cf. the author's *Philosophie der Raum-Zeit-Lehre* (Berlin: De Gruyter, 1928), § 12.

which had been previously neglected. In its further development, however, the tendency has largely trespassed beyond its proper boundaries by highly exaggerating the part occupied by decisions in knowledge. The relations between different decisions were overlooked, and the task of reducing arbitrariness to a minimum by showing the logical interconnections between the arbitrary decisions was forgotten. The concept of entailed decisions, therefore, may be regarded as a dam erected against extreme conventionalism; it allows us to separate the arbitrary part of the system of knowledge from its substantial content, to distinguish the subjective and the objective part of science. The relations between decisions do not depend on our choice but are prescribed by the rules of logic, or by the laws of nature.

It even turns out that the exposition of entailed decisions settles many quarrels about the choice of decisions. Certain basic decisions enjoy an almost universal assent; and, if we succeed in showing that one of the contended decisions is entailed by such a basic decision, the acceptance of the first decision will be secured. Basic decisions of such a kind are, for instance, the principle that things of the same kind shall receive the same names, or the principle that science is to furnish methods for foreseeing the future as well as possible (a demand which will be accepted even if science is also charged with other tasks). I will not say that these basic decisions must be assumed and retained in every development of science; what I want to say is only that these decisions are actually maintained by most people and that many quarrels about decisions are caused only by not seeing the implication which leads from the basic decisions to the decision in question.

The objective part of knowledge, however, may be freed from volitional elements by the method of reduction transforming the advisory task of epistemology into the critical task. We may state the connection in the form of an implication: If you choose this decision, then you are obliged to agree to this statement, or to this other decision. This implication, taken as a whole, is free from volitional elements; it is the form in which the objective part of knowledge finds its expression.

§ 2. Language

It may be questioned if every process of thinking needs language. It is true that most conscious thinking is bound to the language form, although perhaps in a more or less loose way: the laws of style are suspended, and incomplete groups of words are frequently used instead of whole propositions. But there are other types of thought of a more intuitive character which possibly do not contain psychological elements which can be regarded as constituting a language. This is a question which psychologists have not yet brought to a definite solution.

What cannot be questioned, however, is that this is the concern of psychology only and not of epistemology. We pointed out that it is not thinking in its actuality which constitutes the subject matter of epistemology but that it is the rational reconstruction of knowledge which is considered by epistemology. And rationally reconstructed knowledge can only be given in the language form—that needs no further explanation, since it may be taken as a part of the definition of what we call rational reconstruction. So we are entitled to limit ourselves to symbolized thinking, i.e., to thinking formulated in language, when we begin with the analysis of knowledge. If anyone should raise the objection that we leave out by this procedure certain parts of thinking which do not appear in the language form, the objection would betray a misunderstanding of the task of epistemology; for thinking processes enter into knowledge, in our sense of the term, only in so far as they can be replaced by chains of linguistic expressions.

Language, therefore, is the natural form of knowledge. A theory of knowledge must consequently begin with a theory of language. Knowledge is given by symbols—so symbols must be the first object of epistemological inquiry.

What are symbols? It cannot be sufficiently emphasized that symbols are, first of all, physical bodies, like all other physical things. The symbols used in a book consist of areas of ink, whereas the symbols of spoken language consist of sound waves which are as physically real as the areas of ink. The same is true for symbols used in a so-called "symbolic" way, such as flags or crucifixes or certain kinds of salutation by a movement of the hand; they all are physical bodies or processes. So a symbol in its general character does not differ from other physical things.

But, in addition to their physical characteristics, symbols have a property which is generally called their *meaning*. What is this meaning?

This question has occupied philosophers of every historical period and stands in the foreground of contemporary philosophical discussion, so we cannot be expected to give a definite answer at the very beginning of our study. We must start with a provisional answer which may lead our investigation in the right direction. Let us formulate our first answer as follows: *Meaning is a function which symbols acquire by being put into a certain correspondence with facts.*

If "Paul" is the name of a certain man, this symbol will always occur in sentences concerning actions of, or the status of, Paul; or if "north" means a certain relation of a line to the North Pole of the earth, the symbol "north" will occur in connection with the symbols "London" and

"Edinburgh," as for example, in the sentence, "Edinburgh is north of London," because the objects London and Edinburgh are in the relation to the North Pole corresponding to the word "north." So the carbon patch "north" before your eyes has a meaning because it occurs in relation to other carbon patches in such a way that there is a correspondence to physical objects such as towns and the North Pole. Meaning is just this function of the carbon patch acquired by this connection.

One thing has to be considered in order that we may understand this situation. Whether a symbol has the function of meaning does not merely depend on the symbol and the facts in question; it also depends on the use of certain rules called the rules of language. That the order of the town names in the sentence previously cited must be the given one, and not the converse one, is stipulated by a rule of the language, without which the meaning of the word "north" would be incomplete. So it may be said that only the rules of language confer meaning on a symbol. At one time there were found certain stones covered with wedge-formed grooves; it was a long time before men discovered that these grooves have a meaning and were in ancient times the writing of a cultured people, the "cuneiform-writing" of the Assyrians. This discovery comprehends two facts: first, that it is possible to add a system of rules to the grooves on the stones in such a way that they enter into relations with facts of the kind occurring in human history; second, that these rules were used by the Assyrians and that the grooves were produced by them. This second discovery is of great importance to history, but to logic the first discovery is the important one. To confer the name of symbols upon certain physical entities it is sufficient that rules can be added to them in such a way that correspondence to facts arises; it is not necessary that the symbols be created and used by man. It sometimes happens that large stones decay, through atmospheric action, in such a way that they assume the form of certain words; these words have a meaning, although they were not made by men. But the case is still special in so far as these symbols correspond to the rules of ordinary language. It might also happen that forms, obtained by natural processes, would convey European history to us if a certain new system of rules were added—although that does not seem to be very probable. There would still be the question of whether we could find these rules. But very frequently we invent new systems of rules for certain special purposes for which special symbols are needed. The signposts and lights in use for the regulation of motor traffic form a system of symbols different from ordinary language in symbols and rules. The system of rules is not a closed class; it is continuously being enlarged according to the requirements of life. We must therefore distinguish between known or unknown symbolic characters, between actual and virtual symbols. The first are the only important ones, since only actual symbols are employed, and therefore the word "symbol" is used in the sense of "actual symbol" or "symbol in use." It is obvious that a symbol acquires this character not by inner qualities but by the rules of language and that any physical thing may acquire the function of a symbol if it fulfils certain given rules of language, or if suitable rules are established.

§ 3. The three predicates of propositions

After this characterization of language in its general aspect, we must now proceed to a view of the internal structure of language.

The first salient feature we observe here is that symbols follow one another in a linear arrangement, given by the one-dimensional character of speech as a process in time. But this series of symbols—and this is the second conspicuous feature—is not of uniform flow; it is divided into certain groups, each forming a unity, called propositions. Language has thus an atomistic character. Like the atoms of physics, the atoms of language contain subdivisions: propositions consist of words, and words of letters. The proposition is the most important unity and really performs the function of the atom: as any piece of matter must consist of a whole number of atoms, so any speech must consist of a whole number of propositions; "half-propositions" do not occur. We may add that the minimum length of a speech is one proposition.

We express this fact by saying that meaning is a function of a proposition as a whole. Indeed, if we speak of the meaning of a word, this is possible only because the word occurs within propositions; meaning is transferred to the word by the proposition. We see this by the fact that groups of isolated words have no meaning; to utter the words "tree house intentionally and" means nothing. Only because these words habitually occur in meaningful sentences, do we attach to them that property which we call their meaning; but it would be more correct to call that property "capacity for occurring within meaningful sentences." We shall abbreviate this term to "symbolic character" and reserve the term "meaning" for propositions as a whole. Instead of the term "symbolic character" we shall also use the term "sense"; according to this terminology, words have *sense*, and propositions have *meaning*. We shall also say that meaning is a predicate of propositions.

The origin of this unique propositional form arises from a second predicate which also belongs to propositions only and not to words. This is the character of being true or false. We call this predicate the *truth-value* of the proposition. A word is neither true nor false; these concepts are not applicable to a word. It is only an apparent exception if occasionally the use of words contradicts this rule. When children learn to talk, it may happen that they point to a table, utter the word "table," and receive the confirmation "yes." But in this case the word "table" is only an abbreviation for the sentence, "This is a table," and what is confirmed by "yes" is this sentence. (The word "yes" in itself is a sentence, meaning, "The sentence stated before is true.") Analogous cases occur in a conversation with a foreigner whose knowledge of a language is rather incomplete. But, strictly speaking, a conversation consists of sentences.

The atomic sentences which form the elements of speech may be combined in different ways. The operations of combination are enumerated by logic; they are expressed by such words as "and," "or," "implies," etc. By these operations some atomic propositions may be closely connected; in this case, we may speak of molecular propositions.[4]

> Macbeth shall never vanquish'd be until
> Great Birnam wood to high Dunsinane hill
> Shall come against him.

The apparition states here, to inform Macbeth, a molecular proposition. It is one of the rules of language that in such a case the speaker wants to maintain only the truth of the whole molecular proposition, leaving open the question of the truth of the clauses; so Macbeth is right in inferring that the atomic proposition concerning the strange removal of the wood is not maintained by the apparition and that the implication asserted will not affect him. It is a

[4] The words "sentence" and "statement" are also in use. But this distinction being of little importance and rather vague, we shall make no distinction between "propositions" and "sentences" and "statements."

bad habit of all oracles that they make use in this way of the liberalism of logic, which allows the expression of propositions without their assertion, only to deceive a man in respect to a future fact which their superhuman eyes already see.

There are various ways in which language expresses this intention to leave the question of truth open. As for implication, this renunciation is usually expressed by the use of the particle "if," or "in case," whereas the particle "when" expresses the same implication with the additional condition that the premise will be fulfilled at a certain time. "If Peter comes, I shall give him the book" differs from "When Peter comes, I shall give him the book" in this respect; only in the second case is the first clause asserted separately, so that we may infer here that Peter will come. What is left open by "when" is only the time of the realization. The particle "until" used by the apparition is not quite clear, and, if Macbeth had been a logician, he might have asked the crowned child if she could repeat her molecular proposition by saying "when" instead of "until," after putting the first clause into the positive form. Another way of showing that the proposition is not maintained as true is by the use of the interrogative form. To put a question means to utter a sentence without stating its truth or its falsehood, but with the wish to hear the opinion of another man about it. Grammatically the interrogative form is expressed by the inversion of subject and predicate; some languages have a special particle for this purpose which they add to the unchanged proposition, like the Latin *ne* or the Turkish *mi*. On the other hand, a molecular sentence, running from a full stop to the next full stop, is maintained as true.

There is a third predicate of propositions which must be mentioned in this context. Only a small proportion of the propositions occurring in speech are of such a type that we know their truth-value; for most propositions the truth-value has not yet been determined at the moment when they are uttered. It is the difference between verified and unverified sentences of which we must now speak. To the class of unverified sentences belong, in the first place, all propositions concerning future events. These are not only propositions concerning matters of importance which cannot be thoroughly analyzed, like questions regarding our personal position in life, or questions concerning political events; the greater part of such propositions concern rather insignificant events, like tomorrow's weather, or the departure of a tram, or the butcher's sending the meat for dinner. Though all these propositions are not yet verified, they do not appear in speech without any determination of their truth-value; we utter them with the expression of a certain opinion concerning their truth. Some of them are rather certain, like those concerning the sun's rising tomorrow, or the departure of trains; others are less certain, e.g., if they concern the weather, or the coming of a tradesman who has been summoned; others are very uncertain, like propositions promising you a well-paid position if you follow the instructions of a certain advertisement. Such propositions possess for us a determinate *weight* which takes the place of the unknown truth-value; but while the truth-value is a property capable of only two values, the positive and the negative one, the weight is a quantity in continuous scale running from the utmost uncertainty through intermediate degrees of reliability to the highest certainty. The exact measure of the degree of reliability, or weight, is probability; but in daily life we use instead appraisals which are classified in different steps, not sharply demarcated. Words such as "unlikely," "likely," "probable," "sure," and "certain" mark these steps.

Weight, therefore, is the third predicate of propositions. It is in a certain contrast to the second predicate, truth-value, in so far as only one of these two predicates is used. If we know the truth or falsehood of a proposition, we need not apply the concepts of probability; but, if we do not know this, a weight is demanded. The determination of the weight is a substitute for verification, but an indispensable one, since we cannot renounce forming an opinion about unverified sentences. This determination is based, of course, on formerly verified sentences; but the concept of weight applies to unverified sentences. Thus in the system of propositional weights we construct a bridge from the known to the unknown. It will be one of our tasks to analyze the structure of this bridge, to look for the bridging principle which enables us to determine the degree of propositional weight and to ask for its justification. For the moment, however, we shall be content to point out that there is a weight ascribed to unverified sentences, in science as well as in daily life. To develop the theory of weight, which shall turn out to be identical with the theory of probability, is one of the aims of our inquiry. The theory of propositions as two-valued entities was constructed by philosophers in ancient times and has been called logic, while the theory of probability has been developed by mathematicians only in the last few centuries. We shall see, however, that this theory may be developed in a form analogous to logic, that a theory of propositions as entities with a degree of probability may be put by the side of the theory of propositions as two-valued entities, and that this probability logic may be considered as a generalization of ordinary logic. Although this is to be developed only in the fifth chapter of our book, we may be allowed to anticipate the result and to identify weight and probability.

An appraisal of weight is needed particularly when we

want to make use of propositions as a basis for actions. Every action presupposes a certain knowledge of future events and is therefore based on the weight of propositions which have not yet been verified. Actions—unless they be nothing but a play of muscles—are processes intentionally started by men in the pursuit of certain purposes. Of course the purpose is a matter of volitional decision and not of truth or falsity; but whether the inaugurated processes are adapted to attain the purpose is a matter of truth or falsity. This aptness of means must be known before their verification and hence can be based only on the weight of a sentence. If we want to climb up a snow-covered mountain, that of course is our personal decision; and, if anybody does not like it, he may decide against the climb. But that our feet will sink down in the snow when we step on it; that, on the contrary, planks of two meters length will carry our feet; and that we shall slide down the slopes with them almost as quickly and lightly as a bird in the air—this is to be stated in a proposition which, fortunately, possesses a high weight, if our legs are sufficiently trained. Without knowing this it would be rather imprudent to attempt a realization of our desires to get up the snowy slopes. The same situation holds for any other action, whether it concerns the most essential or the most insignificant matter in our lives. If you have to decide whether you will take a certain medicine, your decision will depend on two things: on whether you want to recover your health and on whether taking the medicine is a means appropriate to this end. If you have to decide the choice of a profession, your decision will depend on your personal desires as to shaping your life and on the question whether the profession intended will involve the satisfaction of these desires. Every action presupposes both a volitional decision and a certain kind of knowledge concerning future events which cannot be furnished by a verified sentence but only by a sentence with an appraised weight.

It may be that the physical conditions involved are similar to former ones and that analogous sentences have been verified before; but the very sentence in question must concern a future event and, therefore, has not yet been verified. It may be true that every day at nine o'clock I found the train at the station and that it took me to my place of work; but, if I want to take it this morning I must know if the same will be true today. A determination of the weight, therefore, is not restricted to occasional predictions of wide bearing which cannot be based on similar antecedents; it is needed as well for the hundreds of insignificant predictions of everyday life.

In the examples given the unverified sentence concerns a future event; in such cases the weight may be considered as the predictional value of the sentence, i.e., as its value in so far as its quality as a prediction is concerned. The concept of weight, however, is not restricted to future events; it applies to past events as well and is, in so far, of a wider extension. The facts of history are not always verified, and some of them possess only a moderate weight. Whether Julius Caesar was in Britain is not certain and can only be stated with a degree of probability. The "facts" of geology and of archeology are rather doubtful as compared with facts of modern history; but even in modern history there are uncertain statements. In daily life uncertain statements concerning the past also occur and may even be important for actions. Did my friend post my letter to the bookseller yesterday so that I may expect the book to arrive tomorrow? There are friends for whom this proposition possesses a rather low weight.

This example shows a close connection between the

weights of propositions concerning past events and predictions: their weights enter into the calculations of predictional values of future events which are in causal connection with the past event. This is an important relation; it is to play a role in the logical theory of weights. We may therefore apply the name "predictional value" to the weights both of future and of past events and distinguish the two subcases as direct and indirect predictional values, if such a distinction is necessary. In this interpretation predictional value is a predicate of propositions of every type.

There is one apparent difference between truth-value and weight. Whether a sentence is true depends on the sentence alone, or rather on the facts concerned. The weight, on the contrary, is conferred upon a sentence by the state of our knowledge and may therefore vary according to a change in knowledge. That Julius Caesar was in Britain is either true or false; but the probability of our statement about this depends on what we know from historians and may be altered by further discoveries of old manuscripts. That there will be a world-war next year is either true or false; if we have only a certain probability for the proposition, this is simply due to the mediocre state of sociological prediction, and perhaps some day a more scientific sociology will give better forecasts of sociological weather. Truth-value, therefore, is an absolute predicate of propositions, and weight a relative predicate.

To summarize the results of our inquiry into the general features of language, as far as we have advanced, let us put together the following points. Language is built up of certain physical things, called symbols because they have a meaning. Meaning is a certain correspondence of these physical things to other physical things; this correspondence is established by certain rules, called the rules of language. Symbols do not form a continuous series but are grouped in an atomistic structure: the basic elements of language are propositions. So meaning becomes a predicate of propositions. There are, in addition, two other predicates of propositions: their truth-value, i.e., their being true or false, and their predictional value or weight, i.e., a substitute for their truth-value as long as this is unknown. This triplet of predicates represents those properties of propositions on which logical inquiry is to be based.

§ 4. The language of chess as an example, and the two principles of the truth theory of meaning

We shall now illustrate our theory of language by an example. This example allows a very simple form of language and will therefore show in a very clear way the three predicates of propositions. We shall also use this example to make a further advance in the theory of the three predicates.

We take the game of chess and the well-known rules in use for the notation of the positions, pieces, and moves. This notation is based on a system of two-dimensional co-ordinates containing the letters *a*, *b*, *c*, ... , *h*, for one dimension, and the numbers 1, 2, ... , 8, for the other one; the pieces are indicated usually by the initials of their names. A set of symbols

Kt c 3

represents a sentence; it says, "There is a knight on the square of co-ordinates *c* and 3." Similarly, the set of symbols

Kt c 3—*e* 4

describes a move; it reads, "The knight is moved from the square *c* 3 to *e* 4."

Now let us raise the question of the application of the first two predicates of propositions, meaning and truth-value. The simplicity of our example permits us to discover a close connection between these two predicates: the given sentences of our language have a meaning because they are verifiable as true or false. Indeed, that we accept the set of symbols "*Kt c* 3" as a sentence is only due to the fact that we may control its truth. "*Kt c* 3" would remain a sentence in our language even were there no knight on *c* 3; it would then be a false sentence, but still a sentence. On the other hand, a group of symbols

Kt c g

would be meaningless because it cannot be determined as true or false. Therefore we would not call it a proposition; it would be a group of signs without meaning. A meaningless set of signs is to be recognized by the fact that the addition of the negation sign does not transform it into a true sentence. Let us apply the sign ∼ for negation; then the set

∼ *Kt c g*

is as meaningless as the foregoing one. A false sentence, however, is changed into a true one by adding the negation sign. So, if there is no knight on the square *c* 3, the set of symbols

∼ *Kt c* 3

would be a true sentence.

These reflections are of importance because they show a relation between meaning and verifiability. The concept of truth appears as the primary concept to which the concept of meaning can be reduced; a proposition has meaning because it is verifiable, and it is meaningless in case it is not verifiable.

This relation between meaning and verifiability has been pointed out by positivism and pragmatism. We will not enter for the present into a discussion of these ideas; we want to present these ideas before criticizing them. Let us call this theory the *truth theory of meaning*. We shall summarize it in the form of two principles.

First principle of the truth theory of meaning: a proposition has meaning if, and only if, it is verifiable as true or false.

By this stipulation, the two terms "having meaning" and "being verifiable" become equivalent. But, although this is a far-reaching determination of the concept of meaning, it is not a sufficient one. If we know that a proposition is verifiable, we know that it has meaning; but we do not yet know what meaning it has. This is not altered even if we know what truth-value the proposition has. The meaning of a sentence is not determined by its truth-value; i.e., the meaning is not known if the truth-value is given, nor is the meaning changed if the truth-value is changed. We need, therefore, another determination which concerns the content of meaning. This intension of a proposition is not an additional property which we must give separately; the intension is given with the proposition. But there is a formal restriction which we must add, by definition, concerning the intension, and without which the intension would not be fixed. This additional definition is performed by means of the concept of "the same meaning." All sentences have meaning; but they do not all have the same meaning. The individual separation of different meanings is achieved if we add a principle which determines the same meaning.

To introduce this concept we must alter our chess language in a certain way. Our language is as yet very rigid, i.e., built up on very rigorous prescriptions; we shall now introduce certain mitigations. We may admit a change of the order of letters and numbers: the capital designating the piece may be put at the end; an arrow may be used instead of the dash, etc. Then we can express the same meaning by different sentences; thus the two sentences

Kt c 3—*e* 4
c 3 *Kt*→4 *e*

have the same meaning. Why do we speak here of the same meaning? A *necessary* criterion for the same meaning can easily be given: the sentences must be connected in such a way that, if any observation makes one sentence true, the other is also made true, and, if it makes one sentence false, the other is also made false. It is held by positivists that this is also a *sufficient* criterion. We formulate, therefore, the

Second principle of the truth theory of meaning: two sentences have the same meaning if they obtain the same determination as true or false by every possible observation.

Let us turn now to the question of truth. When do we call a sentence true? We demand in this case that the symbols should be in a certain correspondence to their objects; the nature of this correspondence is prescribed by the rules of language. If we examine the sentence *Kt c* 3, we look to that square which has the co-ordinates *c* and 3; and, if there is a knight at this place, the sentence is true. Verification, therefore, is an act of comparison between the objects and the symbols. It is, however, not a "naïve comparison," such as a comparison which would demand a certain similarity between objects and symbols. It is an "intellectual comparison"—a comparison in which we must apply the rules of language, understanding their contents. We must know for this comparison that the capital denotes the piece, that the letter co-ordinate denotes the column, etc. So the comparison is, in itself, an act of thought. What it deals with, however, is not an imaginary "content" of the symbols but the symbols themselves, as physical entities. The ink marks "*Kt c* 3" stand in a certain relation to the pieces on the chessboard; therefore these marks form a true sentence. Truth, therefore, is a physical property of physical things, called symbols; it consists in a relation between these things, the symbols, and other things, the objects.

It is important that such a physical theory of truth can be given. We need not split the proposition into its "mental meaning" and into its "physical expression," as idealistic philosophers do, and attach truth to the "mental meaning" only. Truth is not a function of meaning but of the physical signs; conversely, meaning is a function of truth, as we noted before. The origin of the idealistic theory of truth may be sought in the fact that a judgment about truth presupposes thinking; but it does not concern thinking. The statement, "The proposition *a* is true," concerns a physical fact, which consists in a correspondence of the set of signs included in *a*, and certain physical objects.

Let us now ask about the third predicate of propositions within our language. We always meet predictional values when actions are in question; so they must appear when the game of chess is actually played. Indeed, the players of the game are continuously in a situation demanding the determination of a weight. They want to move their pieces in such a way as to attain a certain arrangement of the pieces on the board, called "mate"; and to reach that end they must foresee the moves of the opponent. So each player assigns weights to propositions expressing future moves of his opponent, and it is just the quality of a good player to find good weights, i.e., to regard as likely those moves of the opponent which afterward occur. This illustration corresponds to our exposition of the concept of weight: we see that the weight becomes superfluous if a verification is attained but that it is indispensable as long as a verification is not at hand. A player who used only meaning and truth as predicates of his chess propositions would never win the game; when the unknown becomes known to him, it is too late for interference. The predictional value is the bridge between the known and the unknown; that is why it is the basis of action.

Although predictional values are used by everyone, it is very difficult to clarify how they are calculated. In this respect, the determination of the weight of a proposition differs greatly from that of truth. We showed that, for our language, truth could be defined in a relatively simple way. We cannot do the same for the weight. The weight of the future moves is not a question of the physical state of the pieces alone, but it includes considerations about the psychical states of the player. This case is too complicated, therefore, to serve as an example for the development of the theory of predictional values. As we previously said, we shall postpone this development to a later part of our inquiry. Until then, let us regard the possibility of the determination of a weight as a given fact.

§ 5. Extension of the physical theory of truth to observation propositions of ordinary language

The truth theory of meaning is based on the assumption that propositions can be verified as true or false. We developed this theory, therefore, for an example in which the question of verifiability can easily be settled. Propositions of ordinary language, however, are of many different types, and it may be questioned, at least for some of these types, if verification is possible at all. If we want to extend the truth theory of meaning and the physical theory of truth to ordinary language, it will be reasonable to begin with a type of proposition for which verification contains no difficulties.

This rather simple type of proposition is given by sentences of the kind, "There is a table," "This steamer has two funnels," or "The thermometer indicates 15° centigrade." We shall call them *observation propositions* because they concern facts accessible to direct observation—in the current sense of this word. Later on this question will be more precisely examined; it will be shown that to speak of direct verification for these propositions presupposes a certain idealization of the actual conditions. However, it is a good method to begin with a certain approximation to the actual situation and not with the problem of knowledge in all its complexity; for the present we shall start therefore with the presupposition that for observation sentences absolute verification is possible, and we shall maintain this presupposition throughout the present chapter of our inquiry.

We begin with the question of the physical theory of truth and shall postpone the problem of meaning to the following section. This order of the inquiry is dictated by the result of the preceding section, which showed that meaning is a function of truth; so we had better begin with the question of truth.

We can indeed apply our idea that truth is a correspondence between symbols and facts established by the rules of language; but this correspondence is not always easily seen. Only to the extent to which terms occur which denote physical objects is the correspondence obvious. This is evident from the method in which such terms are defined. For this purpose, we might imagine a "dictionary" which gives on one side the words, on the other side samples of the real things, so that this dictionary would resemble a collection of specimens, like a zoo, rather than a book. It is more difficult to establish the correspondence for logical terms such as numbers. We mentioned the example, "This steamer has two funnels." As for the terms "steamer" and "funnel," the corresponding objects will be found in our collection of specimens—but what of "two"? In such a case, we must look for the definition of the term and substitute it in place of the term. This is a rather complicated matter; but modern logic shows in principle how to perform it. We cannot enter here into a detailed description and can only summarize the method developed in textbooks of logistic. It is shown that a sentence containing "two" has to be transformed into an "existence proposition" containing the variables x and y; and, if we introduce this definition into our original sentence concerning the steamer, we shall finally find a correspondence between the funnels and these symbols y and x. So the term "two" is also reduced to a correspondence.

There is still the term "has." This is a propositional function expressing possession. Propositional functions of such a simple type may be imagined as contained in our collection of specimens. They are relations, and relations are given there by examples which represent them. So the relation "possession" might be expressed, say, by a man wearing a hat, a child holding an apple, a church having a tower, etc. This method of definition is not so stupid as it at first appears. It corresponds to the actual way in which the meaning of words is learned by a child. Children learn to talk by hearing words in immediate connection with the things or facts to which they belong; and they learn to understand the word "has" because this word is used on such occasions as those described. Our collection of specimens corresponds to the grand zoölogical garden of life through which children are guided by their parents.

We see that the correspondence between the sentence and the fact can be established if the sentence is true. It clearly presupposes the rules of language; but it presupposes more: it requires thought. The judgment, "The sentence is true," cannot be performed without understanding the rules of language. This is necessary because any correspondence is a correspondence only with respect to certain rules. To speak of the correspondence between men's bodies and men's suits presupposes a rule of comparison; for there are many points in which suits and men differ entirely. What can be said here is that applying certain rules—in the case of this example, geometrical rules—we find a correspondence between these two kinds of objects. The same is valid for the comparison between symbols and objects, and therefore this comparison needs thought. So the physical theory of truth cannot free us from thought. What is to be thought, however, is not the original sentence *a*, but the sentence, "The sentence *a* is true." It may be admitted that this is a psychological question and that it is perhaps psychologically impossible to separate thinking of *a* and of "*a* is true"; only for a very complicated sentence *a* might this separation be possible. To get rid of this psychological puzzle, we may state our conception in the following way: a sentence of the type, "This proposition is true," concerns a physical fact, namely, a certain relation between the symbols, as physical things, and the objects, as physical things. To give an example: the proposition, "This steamer has two funnels," concerns a physical fact; the proposition *A*, reading, "The proposition, 'This steamer has two funnels,' is true," concerns another physical fact which includes the group of signs, "This steamer has two funnels." That is why we call our theory the physical theory of truth. But this theory does not aim to make thinking superfluous; what it maintains is only that the object of a proposition stating truth is itself a physical object.

The physical theory of truth involves difficulties which can only be solved within a theory of types. One of the puzzles occurring here is the following: if the sentence *a* is true, this implies that the sentence *A*, reading, "The sentence *a* is true," is true also, and vice versa; thus *a* and *A* have the same meaning, according to the second principle of the truth theory of meaning. But the physical theory of truth distinguishes both sentences as concerning different facts. To justify this distinction we have to assume that both sentences are of different types and that the truth theory of meaning applies to sentences of equal type only. The sentence *a* cannot concern a fact comprehending the sentence *a;* that we may infer from *a* to *A* is possible only because the sentence *a* in being put before us shows itself to us and furnishes new material which may be considered in the sentence *A* of a higher level. Reflections of this kind have led Tarski[5] to the strict proof that a theory of truth cannot be given within the language concerned, but demands a language of a higher level; by this analysis some doubts[6] uttered against the physical theory of truth could be dissolved.

§ 6. Extension of the truth theory of meaning to observation propositions of ordinary language

Having shown that observation sentences of ordinary language fit in with the physical theory of truth, we shall try now to extend also the truth theory of meaning to this kind of proposition. This extension demands some preliminary analysis concerning the concepts occurring in the theory of meaning as developed.

[5] A. Tarski, "Der Wahrheitsbegriff in den formalisierten Sprachen," *Studia Philosophica* (Warsaw, 1935); cf. also *Actes du Congrès International de Philosophie Scientifique* (Paris: Hermann & Cie., 1936), Vol. III: *Langage*, containing contributions of A. Tarski and Marja Kokoszynska concerning the same subject. Another contribution of Marja Kokoszynska is to be found in *Erkenntnis*, VI (1936), 143 ff.

[6] C. G. Hempel, "On the Logical Positivist's Theory of Truth," *Analysis*, II, No. 4 (1935), 50.

We begin with the first principle. It states that meaning is tied to verifiability. We said above that we would take for granted the possibility of verification, and we shall continue to maintain this presupposition in the present section. But that is to mean only that we shall put aside objections against the term "verification"; we must, however, now analyze the term "possibility."

Before entering upon this analysis, we have to notice that the possibility which we demand does not concern the assumption in question but only the method of its verification.[7] The assumption itself may be impossible; then the verification will furnish the result that the proposition is false. This is allowable because verification has a neutral meaning for us: it signifies determination as true or false. So the proposition, "Hercules is able to bear the terrestrial globe on his shoulders," is verifiable if there is any Hercules before us raising such pretensions; although we are sure that the realization of his contention is not possible, the verification is possible and will show his contention to be false.

We must ask now what is meant by possibility of verification. The term "possibility" is ambiguous because there are different concepts of possibility; we must therefore add a definition of possibility.

First, there is the concept of *technical* possibility. This concerns facts the realization of which lies within the power of individuals or of groups of men. It is technically possible to build a bridge across the Hudson; to build a bridge across the Channel, from Calais to Dover, is perhaps already technically impossible, and it is surely technically impossible to build a bridge over the Atlantic.

Second, there is the concept of *physical* possibility. It

[7] This has been recently emphasized by Carnap, "Testability and Meaning," *Philosophy of Science*, III (1936), 420.

demands only that the fact in question be conformable to physical laws, regardless of human power. The construction of a bridge across the Atlantic is physically possible. A visit to the moon is physically possible too. But to construct a perpetual-motion machine constantly furnishing energy is physically impossible; and a visit to the sun would be physically impossible, too, because a man would be burned, together with his space ship, before reaching the sun's surface.

Third, there is the concept of *logical* possibility. It demands still less; it demands only that the fact can be imagined or, strictly speaking, that it involve no contradiction. The *perpetuum mobile* and the visit to the sun are logically possible. It would be logically impossible however, to construct a quadrangular circle, or to find a railway without rails. This third concept of possibility is the widest one; it excludes only contradictions.

Let us now apply these concepts to the question of verifiability. It must be kept in mind that these three concepts of possibility are to be applied to the method of verification and not to the fact described by the proposition.

The concept of technical possibility is usually not meant when we talk of the possibility of verification. On the contrary, it is emphasized that the postulate of verifiability leaves a greater liberty to propositions than technical possibility would allow. The statement, "Measured from the bridge across the Atlantic, the difference of the tides would be about ten meters," is taken as verifiable because such a bridge is physically possible; from this bridge we would have only to drop a plumb line to the surface of the water and could measure in this way the level of the water—which ships cannot do because they must follow the rise and fall in sea-level. We shall, therefore, reject technical possibility as a criterion for verifiability.

The concept of physical possibility furnishes a frame wide enough for statements of the given kind; but there are other statements which are excluded by it. To these belong statements concerning a very remote future. That there will be, two hundred years hence, a world similar to that of today cannot be verified by me; so this would be a meaningless proposition if we accept physical possibility for the definition of verifiability. This difficulty might be overcome by a small change in the definition of verifiability; we could content ourselves with the verification performed by any human being and renounce our playing a personal role in the process. But there are other sentences which still would be meaningless. Such would be a sentence concerning the world after the death of the last representative of mankind. Or take a sentence concerning the interior of the sun; that there are forty million degrees of heat in the sun's center cannot be verified because it is physically impossible to introduce an instrument of measurement into the sun's bulk. To this category belong also sentences concerning the atomistic structure of matter. That the electrons revolve in elliptic orbits around the kernel of the atom, that they have a spin, etc., is physically unverifiable in the strict sense of the term. Let us call *physical meaning* the concept of meaning as defined by the demand of physical possibility of verification. Then the given sentences have no physical meaning.

The concept of logical possibility is the widest of the three concepts; applying it to the definition of verifiability, we obtain the concept of *logical meaning*. All the examples given above have logical meaning. A statement about the world two hundred years hence is meaningful, then, because it is not logically impossible that I should live even then, i.e., to suppose this would be no contradiction. And to talk about the world after my death, or after the death of the last man, is meaningful because it is not logically impossible that we should have impressions even after our death. I will not say that this concept of meaning presupposes eternal life; it makes use only of the fact that eternal life is no contradiction, and it abstains, prudently, from any presupposition that there be some chance of its being a reality. Similar reflections hold for the example of measurements in the interior of the sun. I can imagine a thermometer of considerable length put into the sun's center, and the mercury column mounting to a degree marked by the figure four with seven zeros; though I do not think that any physicist will ever attempt to construct such a thermometer, there is no logical contradiction in the conception. It contradicts the laws of physics, to be sure; but physical laws are, in the end, matters of fact and not logical necessities. As for statements concerning the structure of the atom I may imagine myself diminished to such a degree that electrons will appear to have the size of tennis balls; if anybody raised an objection to this, I would be able to answer him that such a presupposition involves no contradiction.

If we are now to make a choice between these two definitions of physical meaning and logical meaning, we must clearly keep in our mind that this is a question for a volitional decision and not a question of truth-character. It would be entirely erroneous to ask: What is the true conception of meaning? or which conception *must* I choose? Such questions would be meaningless because meaning can only be determined by a definition. What we could do would be to propose the acceptance of this decision. There are, however, two questions of truth-character connected with the decision. As we showed in § 1, these are the questions as to the decision actually used in science and as to the entailed decisions of each decision. Let us begin here with the latter; instead of suggesting proposals we prefer the method of erecting logical signposts showing the necessary connections for every possible choice.

We see already from the given examples that both definitions of meaning suffer from grave disadvantages. The conception of physical meaning is too narrow; it excludes many sentences which science and daily life obviously accept as meaningful. The conception of logical meaning is better in this respect; but there is the opposite danger that this conception is too tolerant and may include sentences as meaningful which its adherents do not like to see endorsed within this category.

Such sentences indeed exist. The most important type are sentences including an infinity of observation sentences. Take propositions containing the word "all," referring to an infinite number of arguments; or propositions concerning the limit of the frequency in an infinite series of events, as they occur in statistics. It is no contradiction to imagine an observer of eternal life who counts such a series. But the defenders of the truth theory of meaning have a natural aversion to propositions of this type; and they justify this by insisting that such propositions have no meaning. We see that they presuppose, then, the concept of physical meaning. This concept, on the other hand, seems too narrow; we want to remain in agreement with physics and would not like to be obliged to reject such sentences as those concerning the structure of atoms, or the interior of the sun.

Our analysis, therefore, does not lead to a preference for one of the two conceptions. It leads to a "neither-nor"; or, better, to a "both." Indeed, both conceptions are of a certain value and may be used; what is to be demanded is only a clear statement, in every case, as to which of the two conceptions we have in mind.

This corresponds also to the procedure of actual science. There are many famous examples in modern physics of the application of the concept of physical meaning. Einstein's rejection of absolute simultaneity is of this kind; it is based on the impossibility of signals moving faster than light, and this, of course, is only physical impossibility. Applying instead the concept of logical meaning we can say that absolute simultaneity is meaningful because it can be imagined that there is no limit for increasing the speed of signals. The difference of these two concepts of meaning has been formulated as follows: for our world absolute simultaneity has no meaning, but for another world it might have a meaning. The qualification "for our world" expresses the acknowledgment of physical laws for the definition of the possibility of verification. In the same sense, it is impossible only for our world to observe the interior of the electron, and so propositions concerning this subject are meaningless for our world only. If such a clear terminology is used, ambiguity is avoided, and the two conceptions may both be tolerated.

Let us now proceed to an examination of the second principle of the truth theory of meaning in its application to observation sentences. This principle determines that two given sentences have the same meaning when any possible fact will lead to the same truth-value for both the sentences in question. The bearing of this determination must be considered now.

When we introduced the second principle in the example of the game of chess, the full bearing of the principle could not be recognized because the language in question was very simple and concerned only simple objects. In the language of science, however, the second principle obtains a very wide bearing. It happens frequently that certain sentences appear to have a very different meaning, whereas later examination shows that they are verified by the same observations. An example would be the concept of motion. When we say that the body *A* moves toward the body *B*, we believe that we are stating a fact different from the case in which *B* moves toward *A*. It can be shown, however, that both sentences are verified, respectively, by the same observational facts. Einstein's famous theory of relativity can be conceived as a consequence following from the second positivistic principle of meaning. It is the function of this principle to suppress what we might call the subjective intension of meaning and, instead, to determine meaning in an objective way. It is by the addition of this principle only that the antimetaphysical attitude of positivism is completed, having been inaugurated with the first principle.

Some remarks must be added concerning the term "possibility" within the formulation of the second principle—remarks which make use of our distinctions regarding the definition of possibility.

To avoid contradictions, we use for the second principle the same definition of possibility as for the first. Thus for physical meaning the second principle is to be conceived as prescribing the same meaning to two propositions if it is not physically possible to observe facts which furnish a different verification for the two propositions in question; for logical meaning, accordingly, the equality of meaning is dependent on the logical impossibility of finding different verifications. Our example concerning the relativity of motion corresponds to physical meaning. It is physically impossible to find facts which confirm the statement, "*A* moves toward *B*," and do not confirm the statement, "*B* moves toward *A*"—this is the content of Einstein's principle of relativity. Einstein does not speak of a logical necessity here; on the contrary, he emphasizes the empirical origin of his principle, and it is just the words "physically impossible" in which this empirical origin becomes manifest. Analysis has shown that it is logically possible to imagine facts which distinguish the two sentences in question; so it is logically possible to imagine a world in which the principle of relativity does not hold.[8] The concept of absolute motion, therefore, has logical meaning. It is only for our world that it does not apply.

We do not intend to enter into a more detailed analysis of these questions here. The function of the second principle is dependent on the conception of the first one; we shall, therefore, now continue our exposition of the first principle and enter upon a necessary critique of it.

Our discussion of this principle was not satisfactory. We arrived at two definitions of meaning and showed that both could be tolerated; but our subjective feelings are in favor of one of them, namely, of that definition which demands physical possibility of verification, and which accordingly furnishes the more rigorous concept of meaning. The concept of physical meaning looks sounder than that of logical meaning, and the epistemological progress of physics in recent times is indeed due to emphasizing this conception. Einstein's purification of space-time doctrines, the elucidation of the theory of atoms by the quantum theory, and many other similar clarifications have been carried through by the use of the rigorous concept of physical meaning. The advantage of this concept lies in its healthy appeal for restricting sense to descriptions of practicable operations. We spoke of the concept of technical possibility; if this concept is rejected for the definition of verifiability, it is because it cannot be demarcated sharply and would change with the advance of the technical abilities of mankind. The domain of the technically

[8] Cf. the author's "*Philosophie der Raum-Zeit-Lehre*, § 34.

possible has as its upper limit physical possibility; in this sense, we might say that the decision to adopt physical meaning is the decision as to practicable operations. It would therefore be the aim of epistemology to build up a theory of physics in which all propositions concerning our world were justified by physical meaning and did not need to be supported by the concept of logical meaning.

This postulate is not satisfied by the considerations previously developed. We found that sentences concerning events of the remote future, or concerning the structure of the atom, presuppose logical meaning because they cannot be verified if the laws of physics hold. But though this be true, we have the feeling that such a justification by logical meaning does violence to what we really think. We do not agree that we accept a sentence about the temperature in the interior of the sun only because we can imagine a thermometer which obediently continues to perform its functions in conditions under which all other bodies are vaporized. We do not believe that physical statements concerning the structure of the atom have meaning only because we can imagine our own body diminished to atomic dimensions, watching the movement of the electrons as we watch the sun's rising. There must be something wrong in our theory of meaning; and we will try to discover what it is.

§ 7. The meaning of indirect propositions, and the two principles of the probability theory of meaning

A way out of this difficulty has been indicated by pragmatism and positivism. It consists in introducing a second type of verification, which we will call *indirect verification*. There are propositions which cannot be directly verified, but which can be reduced in a certain way to other propositions capable of direct verification. Let us call propositions of this kind *indirect propositions;* accordingly, observation propositions may be called *direct propositions.*

Using these concepts, we construct a solution in the following way. We retain the demand of physical possibility, thus using the concept of physical meaning alone. But those propositions which turn out to be unverifiable on this definition are no longer considered as observation propositions; they change from direct propositions to indirect propositions. So they acquire an indirect meaning; and the occurrence of such propositions in physics is no longer in contradiction to the postulate of physical meaning.

Before entering into a detailed analysis of this plan, let us add a remark. The question whether or not a proposition is a direct one cannot be answered unambiguously; the answer depends on the definition of meaning. Take our proposition concerning the temperature in the interior of the sun; from the standpoint of logical meaning it is direct, from that of physical meaning it is not. The same holds for the term "observation proposition." This term seems to have a clear meaning; but we find that it depends on the definition of the possibility of observation. To observe the temperature in the interior of the sun, in the same sense as we observe the temperature of our chamber, is logically possible but not physically. So all these categories of sentences have no absolute meaning but vary with the definition of meaning.

Let us now take up the question of indirect verification. The determination of this term is suggested by the method of verification used in the practice of science. The sun's temperature is measured in a very complicated way. Physicists observe the energy contained in light rays of different colors emitted from the sun; and, comparing the obtained distribution to analogous observations on terrestrial light rays, they calculate the temperature of the sun's surface. The regularities presupposed in this measurement are involved in the laws of radiation. After determining the temperature on the surface of the sun, physicists, by rather vague and speculative calculations, arrive finally at the number of forty million degrees for the interior of the sun; these calculations contain a number of physical observations of all kinds, especially those involved in the theory of atoms.

We find that in this way the indirect sentence is reduced to a class of direct sentences. These direct sentences concern electrical and optical instruments of measurement, thermometers, colors, etc., but all are situated on our earth in the physical laboratories, so that no visit to the sun is needed. It is true that there is such a reduction of indirect sentences to direct sentences. What we have to study is the kind of relation between the two categories.

Pragmatists and positivists have made an attempt to clarify this relation. This attempt is based on the supposition that there is an equivalence between the indirect sentence, on one side, and the class of direct sentences, on the other side. The structure of this class of direct sentences may be rather complicated; it is not simply built up in the form of a conjunction of the direct sentences, i.e., a combination by "and," but it may contain disjunctions, negations, implications, etc. This is obvious even in a simple case: for measuring the temperature of our chamber we may use a mercury thermometer, or an alcohol thermometer, etc. This "or" will be transferred into the class of direct propositions equivalent to the statement concerning the temperature of our chamber. Let us denote the aggregate of direct propositions by $[a_1, a_2, \ldots, a_n]$, the indirect proposition by A; then positivism maintains the equivalence

$$A \equiv [a_1, a_2, \ldots, a_n] \tag{1}$$

The sign $\equiv$ denotes equality of truth value, i.e., if one side is true, the other side is true too; and if one side is false, then the other side is also false. Applying now the second principle of the truth theory of meaning, we find that the indirect proposition A has the same meaning as the class of direct propositions.

We shall call this method of determining the meaning of indirect propositions the *principle of retrogression*. According to this principle, the meaning of the indirect proposition is obtained by constructing the observation propositions from which the indirect proposition is inferred; the principle of retrogression maintains that this inference is to be interpreted as an equivalence and that the meaning of the conclusion of the inference is the same as the meaning of the premisses of the inference. The meaning of the indirect proposition is accordingly constructed by a retrogression, i.e., by a process inverse to the procedure of the scientist. The scientist advances from observation propositions to the indirect proposition; the philosopher, for the purpose of interpretation, goes backward from the indirect proposition to its premisses. This is the idea expressed by Wittgenstein in his formula: the meaning of a proposition is the method of its verification.[9] Pragmatists have, at an earlier time, expressed the same idea by calling observation propositions the "cash value" of the indirect proposition.[10]

[9] Although this formula is not verbally contained in Wittgenstein's *Tractatus logico-philosophicus* (London, 1922), it expresses his ideas very adequately and has been used, with this intention, within the "Vienna Circle."

[10] Cf. W. James, *Pragmatism* (New York, 1907), Lecture VI: "How will the truth be realized? What experiences will be different from those which would obtain if the belief were false? What, in short, is the truth's cash-value in experiential terms?" This idea goes back to the pragmatic maxim of C. S. Peirce, first pronounced in 1878: "Consider what effects, that might conceivably have practical bearings, we conceive the object of our conception to have. Then, our conception of these effects is the whole of our conception of the object" (*Collected Papers of C. S. Peirce*, V, Cambridge, Mass., 1934, 1). The logical development of the theory inaugurated by this formula is due mainly to James, Dewey, and Schiller.

This equivalence theory of indirect meaning is of seductive power on account of its simplicity and clearness. If it should hold, the theory of knowledge would acquire a very simple form: all that physics states would be a summary of observation propositions. This has been, indeed, emphasized by positivists. But this theory does not survive more rigorous criticism.

It is not true that the class of direct sentences occurring on the right of the equivalence (1) is a finite one. The equivalence sign $\equiv$ means a double implication, i.e., an implication from left to right and another implication from right to left. Hence the propositions $a_1, a_2, \ldots, a_n$, comprehend the whole series of propositions from which *A* can be inferred and at the same time all propositions which can be inferred from *A*. But this is not a finite class; or, at least, it is a practically infinite class, i.e., a class which never can be exhaustively given to human beings. Take as an example the sentence *A* concerning the temperature of the sun. Among $a_1, a_2, \ldots, a_n$ we have, then, observations concerning radiation of sunbeams and hot bodies, observations concerning spectral lines, etc. It is true that the class of propositions from which we start in order to infer *A* is a finite one, and even a practically finite one; for what we have is always a finite number of propositions. But the class of propositions which we can infer from *A* is not finite. We may infer from *A* that the temperature of a certain body, brought within a short distance *r* from the sun, would be *T* degrees; we cannot perform this experiment because we cannot leave the earth's surface. There is an infinite class of such sentences; by making *r* run through all possible numerical values this class would be infinite. It is therefore a grave mistake to think that the right side of (1) can ever be practically given.

This needs an additional remark. There is one case in

which the infinity of consequences drawn from A would present no difficulties: this would be so if the same consequences could be inferred from the finite set $[a_1, a_2, \ldots, a_n]$. In this case, our knowledge of the set $[a_1, a_2, \ldots, a_n]$ would enable us to assert the whole class of consequences drawn from A; there would be no surplus meaning in A, compared to the set $[a_1, a_2, \ldots, a_n]$. But this is obviously not the case in physics. For physical propositions the proposition A has a surplus meaning; and the consequences inferred from A cannot be drawn from the set $[a_1, a_2, \ldots, a_n]$. That the temperature at a distance r from the sun has a determinate value T cannot logically be inferred from $[a_1, a_2, a_3, \ldots, a_n]$; it is logically possible that future observation at a place distant r from the sun would furnish a value different from T in spite of the formerly observed set $[a_1, a_2, \ldots, a_n]$. This is due to the independence of empirical observations; there is no logical compulsion that a future observation should correspond to former ones, or to any expected result. It is because the physical statement A includes predictions for future observations that it contains a surplus meaning compared with the set $[a_1, a_2, \ldots, a_n]$; and it is the indeterminateness of the future which baffles the equivalence theory of positivism concerning indirect sentences.

The real connections are of a more complicated character. We start from a finite class of propositions $[a_1, a_2, \ldots, a_n]$; but from this class there is no logical implication to A. What we have is only a *probability implication*.[11] Let us denote the probability implication by the sign $\ni$; then we have to write

$$[a_1, a_2, \ldots, a_n] \ni A \qquad (2)$$

[11] As to the rules of the probability implication, see the author s *Wahrscheinlichkeitslehre* (Leiden: Sijthoff, 1935), § 9.

On the other hand, even the inferences from A to a_1, a_2, ..., a_n, are not absolutely sure; for it may happen that A is true, whereas $a_1, a_2, \ldots, a_n$ are not true—although this is very improbable. So we have also a probability implication, and not a logical implication, from A to $a_1, a_2, \ldots, a_n$:

$$A \ni [a_1, a_2, \ldots, a_n] \tag{3}$$

The logical equivalence is defined by the double implication; let us accordingly introduce a new term for the mutual probability implication and call it *probability connection*. Using the sign $\ominus$ for this relation, we have

$$A \ominus [a_1, a_2, \ldots, a_n] \tag{4}$$

This probability connection takes the place of the equivalence (1).

The rejection of the equivalence (1) was based on the idea that the class of observation sentences which may be co-ordinated with A is not finite. It may be asked now whether there is at least an infinite class of observation sentences such that it is equivalent to A. This question will be examined later (§§ 15–17); for the present it may be sufficient to say that, if there is such an equivalent class, it is infinite.

Now it is true that the control of an infinite set of observation sentences, one after the other, is only physically impossible, not logically impossible. Thus, if we put aside, for a moment, all other difficulties in the determination of the equivalent class and leave the discussion of these for later consideration, we might say that the admission of logical meaning would enable us to reduce an indirect sentence to an equivalent set of observation sentences. But we must realize that with this interpretation of indirect sentences most propositions of physics are endowed with meaning only because it is not logically impossible to count, term after term, an infinite series. I do not think that such reasoning will convince anyone. Nobody would take such a formal possibility into actual consideration; it is not this logical possibility which leads us to accept the indirect sentences as meaningful. Substantiating the equivalence theory of indirect sentences by reference to the logical possibility of controlling an infinite set of observations would be to destroy the connection between rational reconstruction and actual science and would annihilate the very basis of positivism and pragmatism.

This result expresses the definite failure of the truth theory of meaning. It is not possible to maintain the postulate of strict verifiability for indirect sentences; sentences of this kind are not strictly verifiable because they are not equivalent to a finite class of direct sentences. The principle of retrogression does not hold because the inference from the premises to the indirect sentence is not a tautological transformation but a probability inference. We are forced, therefore, to make a decision: either to renounce indirect sentences and consider them as meaningless or to renounce absolute verifiability as the criterion of meaning. The choice, I think, cannot be difficult, as it has been already decided by the practice of science. Science has never renounced indirect sentences; it has shown instead, the way to define meaning by means other than absolute verifiability.

This means is furnished by the predicate of weight. We showed in § 3 that, in all cases in which the truth-value of a proposition is not known, the predictional value takes the place of the truth-value. So it may perform the same function for indirect sentences. The truth theory of meaning, therefore, has to be abandoned and to be replaced by the probability theory of meaning. We formulate the

First principle of the probability theory of meaning: a proposition has meaning if it is possible to determine a weight, i.e., a degree of probability, for the proposition.

For the definition of the "possibility" occurring here we accept physical possibility. It can easily be shown that this is sufficient to grant meaning to all the examples with which we have dealt; we need not introduce logical possibility because those propositions which demanded logical possibility for obtaining meaning within the truth theory receive meaning within the probability theory as indirect propositions. This becomes obvious if we regard such examples as the statement concerning the temperature of the sun. It is physically possible to ascribe a probability to this statement. It is true that in this case we cannot determine the exact degree of probability, but this is due only to technical obstacles. We have at least an appraisal of the probability; this is shown by the fact that physicists accept the statement as fairly reliable and would never agree to statements ascribing to the sun a temperature of, say, some hundreds of degrees only. It will be our task, of course, to discuss this question of the determination of the probability in a more detailed way; and we shall do that later on. For the present, this preliminary remark may suffice.

The second principle of the truth theory of meaning is now replaced by the following one:

Second principle of the probability theory of meaning: two sentences have the same meaning if they obtain the same weight, or degree of probability, by every possible observation.

As before, the concept of possibility occurring here is the same as for the first principle; so it is once more physical possibility which we accept for our definition.

Let us call the meaning defined by these two principles *probability meaning;* the previously developed concept of meaning may then be called *truth meaning*. By the distinction between physical and logical possibility, truth meaning bifurcates into *physical truth meaning* and *logical truth meaning*. It might be asked whether there is the same bifurcation for probability meaning. Such a distinction turns out to be superfluous because the combination of logical possibility with weight does not furnish a concept distinct from logical truth meaning; if it is logically possible to obtain a weight for a sentence, it is also logically possible to obtain a verification. Only physical reasons can exclude verification and at the same time permit the determination of a weight; if we disregard the laws of physics, we are in imagination free from physical experiments and need not distinguish the possibility of a determination of the weight and of verification. Thus logical probability meaning and logical truth meaning are identical. Probability meaning, therefore, is always physical probability meaning. We may therefore drop the addition "physical" and speak simply of probability meaning; both probability meaning and physical truth meaning may be comprehended by the name *physical meaning*.

The probability theory of meaning may be considered as an expansion of the truth theory of physical meaning in which the postulate of verifiability is taken in a wider sense, including the physical possibility of determining either the truth-value or a weight. We shall therefore include both theories under the name *verifiability theory of meaning*. The narrower sense of verification will be expressed by "absolute verification."

The justification of this expansion is given by the fact that this theory, and only this theory, corresponds to the practice of science. When a man of science speaks of the temperature of the sun, he does not take his sentences as meaningful because there is a logical possibility of direct verification but because there is a physical possibility of inferring the temperature of the sun from terrestrial observations. The man of science also knows that this inference is not a logical inference but a probability inference. It may happen that all his premises $a_1, a_2, \ldots, a_n$ are true but that the result A of his inference is false; therefore he can maintain A only with a certain probability.

Some additional remarks must be added. We introduced the concept of "indirect proposition" to obtain meaning for sentences which had none under the presupposition of a certain definition of meaning, but which had meaning under another definition of meaning, being then observation propositions. There are, in addition, other propositions which are in no case observation propositions for any of the definitions of meaning, and which must be conceived as indirect propositions for every theory of meaning. Such are propositions concerning the development of mankind, concerning biological species, concerning the planetary system—in general, sentences the objects of which are so large, or so temporally extended, that a direct view of them is in no case possible. To these propositions belong, in addition, statements concerning abstract matters, such as the spirit of the Renaissance, the egoistic character of a certain person, and the like. All these propositions have to be treated as indirect.

For these propositions also our contention is valid that there is, in general, no logical equivalence between the general or abstract proposition and the aggregate of observation propositions on which they are based. This is obvious from the fact that we are never absolutely sure of the indirect proposition, although the basic propositions may be of the highest certainty. The facts from which we infer the egoistic character of a man may be undoubtedly certain; but that does not exclude our observing at some later date some actions of the man which are not compatible with the hypothesis of egoism. Propositions of this kind demand the same expansion of the concept of meaning as was given before; it is only the probability theory of meaning which can do justice to them, without doing violence to the actual use of such propositions in science or in daily life. So we cannot accept the positivistic interpretation that these propositions are equivalent to a finite set of verifiable propositions; we take them as meaningful only because they possess a certain weight derived from observations.

§ 8. Discussion of the verifiability theory of meaning

We have now to consider some objections which may be raised against the verifiability theory of meaning. Since this term is to include both truth theory and probability theory of meaning, we are speaking here of objections raised against both theories in common; such a common discussion is possible because the probability theory is a continuous expansion of the truth theory of meaning.

The usual objections start from the fact that the concept of meaning is frequently used without special reference to verification. Poets talk of ancient myths, religious men of God and the heavens, scientific men of the possible origin of the world, without being interested in the question of verification. They may agree that in these cases verification lies beyond human power; but they are convinced that in spite of this their ideas at least have meaning. They even see images with the "mind's eye" and feel sure that they have a clear idea of what they intend. Is not this psychological fact a proof against the connection of meaning and verifiability?

To this we must answer that the cases considered are not of a uniform character and must be carefully classified.

There are many cases in which not the verifiability but the truth is to be denied. Stories invented by poets, and old myths, are surely not true; and just on this account they are verifiable, this term denoting only the neutral quality that a determination as true or false is possible. So these cases are not examples of a separation of meaning and verifiability. On the other hand there are cases in which the verifiability is questioned indeed, as in the case of many religious statements which their adherents frequently advance with the pretension that no human knowledge can ever verify their truth.

We are referring here mainly to religious mysticism, which in all times has exercised a great influence upon men, but whose doctrines cannot be measured in the scale of scientific truth. The utterances of religious prophets are frequently of such a kind that strangers do not understand them at all, whereas the believers are raised to the highest exaltation; or, if there is an ordinary sense in the words used, it is maintained by the adherents that this verifiable part of the doctrine is not the essential meaning—that there is a "higher" meaning which has nothing to do with verifiability.

Before entering into an analysis of this conception, we may make a general remark. If we intend to contest the right of mystics to speak of their speech as meaningful, this is not to question the relevance their utterances may have for themselves or for their auditors. It would be a naïve intellectualism to contest the moral and esthetical value which mysticism may have and actually has had in the history of the human spirit. But, if mystic utterances may have significance, this does not imply that they also have signification. Music too has an effect of the highest order on men and may be one of the best means of spiritual and moral education. But we do not speak of the meaning of music. In this case the lack of the property "meaning" is obvious because music does not possess the external forms of language. Mystic utterances, however, show such forms; this is the reason why the emotional and educational character of such utterances may be confounded with what we call "meaning."

It is a matter of fact that language is not always used with the intention of communicating something to other persons. Language may be used for the purpose of influencing persons, of raising in them certain states of feelings which we want to have produced in them; and language may be a good instrument for this, even better sometimes than music, which if not accompanied by speech may have incomplete effects only. A good preacher may raise the feelings of devotion, penitence, contrition, or the impulse for a life according to the moral conceptions of the church by means of his sermon; and the effect of the accompanying chants may be confined to a subordinate role in comparison with his speech. A politician, by means of his speech, can force his opinion upon a meeting even in case rational reflections should refute his views. Colloquial language also is never entirely free from such a suggestive component—be it the suggestion contained in a salesman's speech to a customer, or in a teacher's speech to his pupil, or in the speech of friend to friend. But the *suggestive* function of language must be logically separated from its *communicative* function, i.e., its function of informing other persons about certain facts or relations between facts.

There is still a third function of language which must be distinguished from its communicative function. Language may release us from an inner constraint, may slacken a tensed mind—be it the oppression caused by physical or psychical pains, or the delightful tension of joy, or the nervous constraint of productive situations of a creative mind. The *relaxive* function expresses itself in a whole range of diverse forms—the "Oh" uttered when a needle pricks our finger, a tune whistled to one's self, the verses releasing the emotional tension of a poet. This relaxive function of language is as different from the communicative function as is the suggestive function; it may show relations to the latter in assuming an autosuggestive function, such as in the case of a child's talking loudly in entering a dark chamber alone. We may combine these two functions, the suggestive and the relaxive function, in the term *emotional* functions, indicating that it is the emotional sphere which is concerned, and leaving open the possibility of adding other functions of a similar character.[12]

It is not our task here to point out why emotional functions are so well performed by the use of utterances which at the same time have a communicative character; what interests us is the question of the logical determination of the communicative function. This determination is not free from arbitrariness; but it seems to me that there are two factors indispensable for any such definition if it is to correspond to the use of speech in practical life.

The first is that a communicative function begins only when there are certain rules established for the use of the terms. We spoke of the relaxive function the word "Oh" may have for a person pricked by a needle; now imagine a person sitting in a dentist's chair and receiving the order to indicate any feeling of pain caused by the drill. The "Oh" uttered in such a case—though not losing, happily, its relaxive function—possesses at the same time a communicative function; it communicates to the dentist the fact that his drill has pierced the thin surface of the tooth's

[12] We follow, in the exposition of the different functions of language, ideas developed by Ogden, Bühler, and Carnap.

enamel. This "Oh" is a sentence endowed with meaning; it is so because it is an utterance in correspondence with the rules established by the dentist's order. It is the adaptation to certain rules which transforms an utterance with a relaxive character into one with communicative character, i.e., into a proposition (cf. also § 2).

The rules we speak of are arbitrary within wide limits; but there is one property—and this is the second essential factor we wish to indicate—which we demand if they may be called rules determining a meaning. This property is the occurrence of something such as a truth-value. For this we do not demand absolute truth; our predicate of weight is a sufficient representative of what is to be demanded here. But some such determination must occur; we must be able to assent to, or to deny, a sentence, or at least to assent to it in some degree. There never was, indeed, a theory of meaning which contradicted this postulate. Mystic utterances are set forth by their adherents with such a claim, even with pretensions of an extremely high degree of truth; for mystics talk of the absolute truth of their doctrines. This is just why they distinguish their discourse from emotional stimuli such as music. Music, though it may be suggestive, exciting, powerful, is not true, whereas the speech of a mystic pretends to be true, absolutely true.

If the verifiability theory of meaning is then questioned by philosophers who want to support mysticism, or any kind of "nonphysical" truth, it is not the predicate of truth-value which is attacked by them. What they attack, instead, is the verifiability of such propositions; they do not acknowledge that it must always be possible to determine the truth-value by observational methods. The religious man maintains his statements concerning God, the Judgment Day, etc., as true but admits that there is no possibility of proving their truth empirically. It is, therefore, the difference between existence of the truth-value and empirical determinability of a truth-value which constitutes the subject of every discussion concerning the verifiability theory of meaning.

With this formulation the problem of the definition of meaning acquires a more definite form. We have distinguished three kinds of meaning which we called *physical truth meaning*, *probability meaning*, and *logical meaning*. Let us introduce a fourth term for the kind of meaning presupposed in religious or mystic speech; let us call it *super-empirical meaning*. The adherents of this kind of meaning, we said, do not contradict the idea that a statement is to be true or false; they do not admit, however, that the usual methods of empirical science are the only means to determine a truth-value. They oppose, therefore, super-empirical meaning to empirical meaning, combining in the latter term the three other kinds of meaning mentioned. The logical order of the four kinds of meaning may be indicated by the diagram in Figure 1; if we consider the classes of propositions admitted as meaningful by each of these definitions, their extensions form domains which include or are included in one another.

We must now analyze the question as to the choice between empirical and super-empirical meaning. This question, we must admit, cannot be raised in the form of whether we are forbidden or allowed to decide for super-empirical meaning. We have made clear that the question of meaning is not a matter of truth-character but of definition and, therefore, a volitional decision; thus a question as to our being forbidden or allowed one usage or another cannot be raised. As we pointed out in § 1, instead there are two questions of truth-character connected with the decision. They concern the decision actually used in science, and what we call the "entailed decisions." The first of these questions does not interest us at the moment; we wish to make a choice, to decide on a definition. It is therefore the second question, the question of the entailed decisions, which we have to raise; it is only in answering

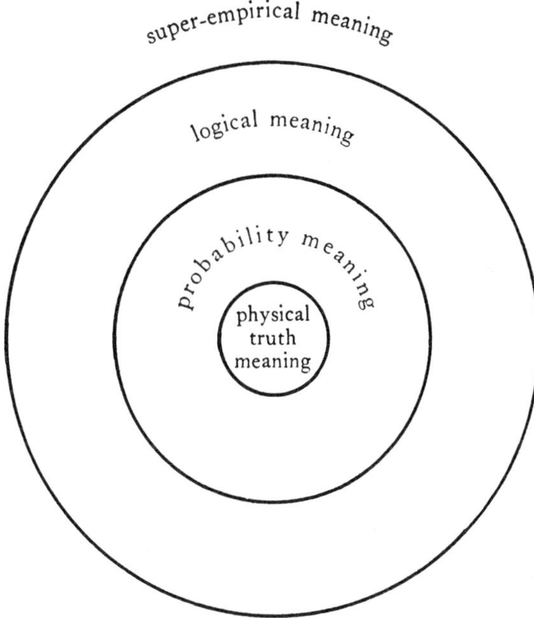

FIG. 1.—The different kinds of meaning

this question that we shall find a basis for settling the question of the connection of meaning and verifiability.

Positivists have advanced the idea that statements which have super-empirical meaning are empty; we pretend, it is said, to mean something, but we do not mean anything. I do not think that this is a clear refutation. It is difficult to convince a person that his words mean nothing; this is because the acknowledgment of this contention depends on the definition of the terms "meaning something" and "meaning nothing." Under what conditions is a statement empty? If this is the case when a statement is not verifiable, then of course super-empirical meaning is empty; but how could we convince a person that he should accept this definition of emptiness? Arguments of this kind are *argumenta ad hominem;* they may persuade certain persons, but they do not clarify the problem.

The question of the entailed decisions is clear and unambiguous. It leads to an indubitable distinction between the issues relevant to the decisions in favor of empirical or super-empirical meaning.

To carry through this analysis, we must first introduce a classification of super-empirical statements. Let us group into one class all those statements for which it is maintained that we have no means at all for knowing their truth-value; into the other class we put all those statements the truth-value of which is known, but by super-empirical methods.

As to the first of these two classes, we may now indicate a property which distinguishes it from empirical statements. This property concerns the applicability of such statements for the purpose of actions. If we want to make use of a statement in the pursuit of a certain action, we must know its truth-value, or at least, its weight. We do not intend to say that statements of known truth-value are a sufficient basis of actions; we explained previously (§ 3) that an action always presupposes a volitional decision concerning an aim. But besides this fixing of the aim, we need a certain knowledge, i.e., statements with a truth-character, to attain the aim; they indicate the way of the realization. Now it is obvious that this function can only be performed by statements the truth-value or weight of which is known. It follows that the statements of our first class of super-empirical statements never can be used as bases of actions.

Let us proceed now to the second class. It seems that for these statements an inapplicability for actions cannot be maintained. Religious belief has been historically the source of many actions, and even of actions of the greatest import. The ideas that the world is a creation of God, that God is omnipotent and omnipresent, that there is a life after death, etc., have played a great role in human history. It is admitted that empirical proofs of these statements cannot be given; but there were at all times adherents of such ideas so highly convinced of their super-empirical truth that they did not hesitate to lead wars, to kill people, or to sacrifice their own lives, when the acknowledgment of such statements demanded it.

To analyze this problem, we must first point out that not all religious statements are divested of empirical meaning. The statement of life after death involves future experiences similar to those we have in ordinary life; if we must contest its physical truth meaning, we cannot deny its logical meaning. Such statements may become bases of actions if they are supposed to be true; for, if a statement is to become a basis of actions, it is sufficient if we think it to be true. If the primitive man puts jars with food and water into the tombs of his friends, this action is correctly derived from his belief that his friends will continue to live after death. In such a case, our inquiry has to take another direction; we have to ask whether there are methods for discovering the truth-value of statements having logical meaning. The answer is given in the discussion of scientific methods; it is shown there that this is possible only if there are at least probability inferences to such statements, that is, if they belong to that part of logical meaning which coincides with probability meaning. And if so, we cannot admit that there is a super-empirical determination of their weight different from a determination by empirical methods. For the other part, for the domain of merely logical meaning, there is no possibility of determining the truth-value, or weight; it follows that the inferences derived from such statements and leading to actions are false—that they are simply a false substantiation of actions. This does not mean that the statement is false but that the substantiation is false; the truth-value of the statement is unknown and precisely for this reason no inferences concerning actions can be deduced from it. The status of this kind of statement, therefore, is settled by reflections belonging to the discussion of science, and we may abandon further discussion here.

What is of a greater significance to us is the discussion of genuinely super-empirical statements—statements which have not even logical meaning. It is the second class of these statements, those which are considered as true, which we must now consider.

Let us ask for the relation of such statements to actions. It seems that such statements may be applied to actions; we cannot demonstrate, as for statements having logical meanings, that their truth-value must necessarily remain unknown—we cannot because they are not submitted to the methods of probability calculations. If some people believe that the cat is a divine animal, they do not claim to be able to prove this empirically; in spite of that, such a belief may determine their actions. It may, for instance, prevent them from killing cats. In this case, a super-empirical statement may become relevant for actions.

To analyze this problem, let us proceed to a closer analysis of the given example. We may first ask our cat worshiper for the reasons of his belief. He may answer that there are some indications of divine character in cats, such as the sparkling of their eyes, but that a full proof cannot be given empirically; he knows directly, he says, about the divine character of cats because they raise in him a certain feeling of awe—in short, he feels the cat's divinity. It is this immediate knowledge which determines him never to kill a cat.

It is not our intention to dissuade our cat worshiper from his belief. What we oppose to his religious conviction is a statement of a very modest type. What he calls a divine animal, we say, may be called by us an animal which raises, in certain people, feelings of awe—in short, an "emotion-producing" animal. To his super-empirical concept "divine" we co-ordinate in this way the empirical concept "emotion-producing"; it is empirical because it is defined by the occurrence of certain psychological reactions in man, belonging to the sphere of observational facts.[13] Our co-ordinated concept is equivalent to his in the following sense: every action which he may derive from his super-empirical meaning may be derived from our co-ordinated empirical meaning as well. His principle, e.g., that divine animals must not be killed, reads with us: emotion-producing animals must not be killed.

Our opponent may object that this equivalence does not hold for him. He observed frequently, he says, that people are persuaded if someone tells them that "divine animals must not be killed"; but the profane words, "emotion-producing animals must not be killed," do not convert them. This may be true; yet it proves nothing but a special suggestive influence attached to the word "divine"—no more. We spoke above about the suggestive function of language; we see now that two propositions which logically determine

[13] We invoke here psychological facts but leave the question as to the character of psychological facts to a later investigation (cf. § 26).

the same consequences may differ as to their suggestive effects. The super-empirical meaning therefore reduces to a surplus suggestive effect; it does not lead us, however, to actions different from those determined by empirical meaning, if the volitional decisions are made in a corresponding way.

We do not forbid anyone to decide for super-empirical meaning; but he cannot rid himself of the consequence that we may co-ordinate to his propositions others of empirical meaning which have the same bearing upon our actions. The "super-empirical content" of the proposition, therefore, is not utilizable, not convertible; super-empirical propositions are like inconvertible papers which we keep in our safe without the possibility of any future realization. This is the result of our critical analysis of the different definitions of meaning, carried through by means of the question of the entailed decisions.

The expediency of this characterization may be questioned by pointing out the fact that there are many verifiable statements, and even statements known as true, which we never use as a basis of actions. This is true; it is due to the fact that our knowledge is much larger than the domain of practically useful sentences. We know that Charlemagne died in 814, or that the moon is at a distance of 238,840 miles from the earth, or that the English language has about 400,000 words; and indeed we make no practical use of this knowledge. But we might do so; and it may happen that some day we shall be placed in some situation which demands the utilization of this knowledge. With regard to Charlemagne, it might happen that a quarrel concerning an inheritance, or the right to bear a certain title, depended on the year of his death; the moon's distance will gain practical importance at the moment when space navigation is rendered practicable, and the size of the vocabulary of the English language has its practical bearing at the moment when a complete English dictionary is to be constructed. I do not say that meaning *is* utility, or that truth *is* utility; I only say that sentences having empirical meaning *may* become useful. Neither do I say that they are true because they may become useful; I say that they may become useful because they are verifiable. It is not the definition of truth or of weight which is to be given here; these concepts are presupposed in the present discussion. It is the definition of meaning which we discuss, and the question whether this term is to be made a function of truth or of weight; we base this decision on the fact that the verifiability definition of meaning leads to a combination of meaning and utilizability, and determines meaningful propositions as those which may be used as the basis of actions.

Is this pragmatism? The answer may be determined by those who have a better knowledge of pragmatism than I have. For the theory developed here it is essential that meaning is not defined in terms of utility but in terms of truth and weight; only the argument for this choice of the definition is furnished by its relation to utilization. This relation is in itself a statement which we maintain as true; it may be seen from this that theories about the combination of meaning and utilizability presuppose the concept of truth and that truth cannot be defined by utilizability. As far as I see, pragmatists did not clarify these rather complex relations. But our conception may perhaps be taken as a further development of ideas which originated in pragmatism. It was the great merit of the founders of pragmatism to have upheld an antimetaphysical theory of meaning at a time when the logical instruments for a theory of knowledge were not yet developed to such a high degree as in our own day.

It is the advantage of our characterization of the verifiability theory of meaning that it does not prescribe the verifiability definition of meaning but that it clarifies this definition together with its entailed decisions. It is the method of the logical signpost which we apply here, leaving the decision to everyone as his personal matter. If we decide, personally, for the verifiability theory, this is because its consequences, the combination of meaning and action, appear to us so important that we do not want to miss them.

We must ask, however, whether the substantiation we give here for empirical meaning applies to each of the three kinds of meaning comprised by us in the concept of empirical meaning. In entering into this inquiry, we shall meet with remarkable results.

We have already pointed out that the domain of merely logical meaning includes propositions which can never be used for action. This is because their truth-value is not accessible to us. Thus this domain turns out to be of a kind similar in this respect to super-empirical meaning; propositions of merely logical meaning as well as super-empirical propositions are inconvertible, are not utilizable for actions.

On the other hand, if we regard physical truth meaning, we find that this definition cannot be justified by utilizability either. In § 3 we discussed the difference between truth and weight, and we showed that truth can only be determined for sentences concerning the past; whereas sentences concerning the future can be ranged only within the scale of weight, their truth-value being unknown to us. We added that this entails a preponderance of weight, in opposition to truth, as soon as the viewpoint of action is introduced; for actions are to be based on statements concerning the future. Statements concerning past events assume importance for actions only in so far as they lead to statements concerning the future, i.e., in so far as they furnish a basis for a determination of the weight of statements. The problem of these inferences to statements concerning the future embraces the problem of induction and will be analyzed later; independently of the result of such analysis, it is obvious that only sentences with an appraised weight furnish the direct basis for actions, not sentences known as true. The argument which we gave in favor of the verifiability theory of meaning—that those sentences which can furnish a basis for actions are to be regarded as meaningful—turns out, therefore, to be an argument in favor of the probability theory of meaning, and to distinguish it from the truth theory. The truth theory is too narrow; it takes as meaningful only a part of the propositions used as a basis of action, and only that part which furnishes the indirect basis, needing in each case the completion by propositions of another class, of the class of sentences with an appraised weight. It would be erroneous to say that these sentences are a possible basis for action only because they will eventually be verified as true or false; for as soon as they are so verified they are no longer a basis for action—the events described in the sentences being then passed and no longer accessible to actions. It is therefore just the predicate of weight which indicates the link between statement and action.

So our analysis leads us to ascribe a unique position to the probability theory of meaning. It is just this theory of meaning which is distinguished by the postulate of a relation between meaning and action. The line of separation in the domain of meaning, as far as it is determined by the postulate of utilizability of statements, cuts through the domain of empirical meaning; it leaves the merely logical meaning on the same side as super-empirical meaning, determining both as comprehending inconvertible statements. On the other side of the line, we find both physical truth meaning and probability meaning; but the first only because it is connected with the second—only because true sentences may lead to sentences having a weight, can they serve as a basis for action. Combining, as in § 7, both physical truth meaning and probability meaning under the name of *physical meaning*, we may say that the domain of physical meaning is the utilizable domain. Therefore it is the probability theory of meaning alone which allows us to satisfy the postulate connecting meaning and utilizability.[14]

This is of importance in respect to a criticism of positivism. Positivists have defended their concept of meaning by insisting that only theirs has meaning; we found that this is an unwarranted absolutism, and that the question of the entailed decisions of the given definition of meaning had to be raised. We tried to show that there is a distinction in favor of a definition which connects meaning with verifiability, but we discover now, on a more exact consideration, that this distinction is opposed to a theory which restricts meaning to absolutely verifiable sentences

[14] Among my former publications concerning the probability theory of meaning, I may mention the following. The idea that empirical propositions are not to be conceived as two-valued entities but are to be dealt with as having a "truth-value" within a continuous scale of probability (a view which demands that they be considered within a probability logic) was first expounded by me at the first congress of "Erkenntnislehre der exakten Wissenschaften" in Prague in 1929 (cf. *Erkenntnis*, I [1930], 170–73). A continuation of these ideas was presented to the following congress, held in Königsberg in 1930 (cf. *ibid.*, II [1931], 156–71). The construction of the probability logic demanded by me has been carried through, in the form of a logistic calculus (including the theory of modalities), in my paper "Wahrscheinlichkeitslogik," *Berichte der Berliner Akademie Wissenschaften* (math.-phys. Kl. [1932]); cf. also my book *Wahrscheinlichkeitslehre*. The two principles of the probability theory of meaning given in § 7 were first formulated in "Logistic Empiricism in Germany and the Present State of Its Problems," *Journal of Philosophy*, XXXIII, No. 6 (March 12, 1936), 147–48 and 154.

—sentences verifiable as true or false only. In our search for tenable arguments in favor of the verifiability theory of meaning, we find therefore that these arguments lead to an expansion of this theory; they should incline the positivist to connect meaning with the wider concept of weight and not with the concept of truth.

Our theory of meaning may therefore be called a further development of positivism, as well as being conceived as a further development of pragmatism. This connection with positivism has a psychological foundation. It seems to me that the psychological motives which led positivists to their theory of meaning are to be sought in the connection between meaning and action and that it was the postulate of utilizability which always stood behind the positivistic theory of meaning, as well as behind the pragmatic theory, where indeed it was explicitly stated. Yet what was overlooked, at least by positivists, was the fact that no true statements concerning the future can ever be attained. This corresponds to the state of epistemology at the time of the foundation of positivism. The probability character of knowledge was not recognized; the laws of physics were taken to be strictly valid for empirical phenomena, and it was tacitly supposed that they furnish statements concerning the future which are to be taken as absolutely true. We read in the books of the older positivists that the object of science is to foresee the future and that this constitutes the very significance of science. This was said, however, without considering the fact that predicting the future presupposes inductions and that the problem of induction must be solved before a theory of meaning can be given which includes the predictive function of science. Although the problem of induction had been unfolded in all its rigor by Hume, its relevance was not seen, and a naïve absolutism concerning future-propositions was joined to the verifiability conception of meaning. But on account of this very combination, the latter conception did not lead to far-reaching restrictions of the content of science.

A more critical attitude was developed in the second phase of positivism—its critical phase. Hume's skeptical objections against induction were accepted, and the failure of any attempt to arrive at a logical solution of induction became more obvious in terms of the pretensions of precision developed in logistics. The impossibility of obtaining certain knowledge about future events was recognized, and this cognizance led, in combination with the postulate of logic as two-valued, to the repudiation of every attempt to interpret scientific propositions as forecasts of future experience. Thus resulted the modern positivistic theory, a strange combination of common-sense elements with a doctrinaire radicalism, which contradicted every unbiased view of the intentions of science. The postulate of absolute verifiability, when pronounced within science, has been mitigated by inconsequent application and therefore could do no harm; but in the hands of philosophers it was exaggerated to a radicalism which questioned the legitimacy of the very aim of science—the prevision of the future. Wittgenstein, the most radical mind among modern positivists, writes: "That the sun will rise to-morrow is a hypothesis; and that means that we do not *know* whether it will rise."[15] He does not realize that there are degrees in the domain of the unknown, such as we have expressed by the predicate of weight. Keeping strictly to the postulate of absolute verifiability, he arrives at the conclusion that nothing can be said about the future.

This does not imply for him that future propositions are meaningless; they have meaning, but their truth-value is

[15] *Op. cit.*, p. 181.

unknown. It indicates, however, that he cannot construct a connection between meaning and action. If we divest his theory of its dogmatical attire, and apply our test of the decisions entailed, we come to the following determination: for Wittgenstein a sentence is meaningful when we can wait for its verification. The stress is on the term "wait for"; we cannot actively utilize the proposition—we can only passively wait for knowledge about it. It is obvious that for this purpose his definition of meaning as verifiability is sufficient. But it is obvious also that in this way an important and healthy tendency of the older positivism has been abandoned—the tendency to combine meaning and action. The decomposing process of analysis has not been accompanied in this case by a constructive process; the possibility of basing meaning on the predicate of weight has been overlooked because a satisfactory interpretation of this predicate could not be developed. The key to a theory of meaning corresponding to the intentions of physics lies in the probability problem. It has been the fate of the positivistic doctrines that they have been driven by logical criticism into an intellectual asceticism which has suppressed all understanding of the "bridging" task of science—the task of constructing a bridge from the known to the unknown, from the past to the future. The cause for this unhealthy doctrinairism is to be found in underestimating the concept of probability. Probability is not an invention made for the sport of gamblers, or for the business of social statistics; it is the essential form of every judgment concerning the future and the representative of truth for any case where absolute truth cannot be obtained.

A further consequence of this lack of insight into the significance of the concept of probability becomes manifest in the erroneous interpretation of the relation between direct and indirect sentences. The principle of retrogression has its origin in mistaking the probability relation between these two kinds of sentences and in replacing it by an equivalence. This principle may therefore be considered as the typical expression of the too narrow logicism which characterizes this form of positivism, of the unwarranted simplification which does violence to the actual structure of science. The positivism of the radical sort cannot be considered as an interpretation of indirect sentences corresponding to the practice of physics.

The more tolerant representatives of positivism recognized this discrepancy between their theory and actual science; and so they looked for an expansion of the narrow definition of meaning previously accepted. Carnap in some recent publications[16] has developed an expansion of the criterion of the meaningful in which the idea of absolute verification is abandoned; he introduces instead the concept of "degree of confirmation," which furnishes a graduated series of propositions, and which is to apply to predictions as well as to propositions concerning past events. This "degree of confirmation" corresponds, in many respects, to our "weight"; with the difference, however, that Carnap doubts whether it is identical with "probability." It seems to me a sign of great progress that with this new theory of Carnap the development of the conceptions of the Vienna Circle turns in a direction leading to a closer connection with physics and to a better approximation to the actual state of knowledge; with this change an old difference between Carnap's conceptions and mine, which was the subject of many a discussion,[17] is considerably reduced. A discussion of Carnap's new conception must,

[16] "Wahrheit und Bewährung," *Actes du Congrès International de Philosophie Scientifique, 1935* (Paris, 1936), IV, 18; "Testability and Meaning," *Philosophy of Science*, III (1936), 420, and *ibid.*, IV (1937), 1.

[17] Cf. the discussion on the congress of Prague, 1929, reported in *Erkenntnis*, I (1930), 268–70.

however, be postponed until he has given some additional information concerning a determination of his "degree of confirmation" and the rules of operating with it. From our point of view, all these questions are answered by the theory of probability, and chapter v will present our answers in detail; but, if the interpretation in terms of probability is not accepted by Carnap, he must develop a theory of his own about degrees of confirmation. The main difficulty of such a theory will lie in the problem of the application of the degree of confirmation to actions; the problem of induction will arise for Carnap in a new form if the solution of this problem within the frame of a logic of probability, such as developed by me, is not considered as applicable to his interpretation of the "weight" of propositions.

Let us add some words concerning the second principle of the verifiability theory of meaning. As we showed, it is the logical function of this principle to cut off any surplus meaning which might be supposed in a proposition beyond its verifiable content. It performs this function in a very "polite" way: it does not forbid "metaphysical" concepts, like forces, tendencies, essences, and deities, but it states: if there is an equivalent nonmetaphysical proposition, i.e., a proposition which does not use these terms, but which has the same truth-value as the first one for all possible facts, then both propositions have the same meaning. Thus the "metaphysical" proposition is deprived of its pretended surplus meaning and reduced to an equivalent nonmetaphysical proposition. This process of cutting off metaphysical claims was first insisted upon by the nominalists of the Middle Ages. William of Ockham pronounced the principle in the form, "entia non sunt multiplicanda praeter necessitatem," and since that time "Ockham's razor" has been the program of every consequent empiricism or logicism. Leibnitz' "principium identitatis indiscernibilium" and its application to the problems of space and motion, Hume's reduction of causality to an invariable succession in time, and Mach's criticism of the concept of force and of Newton's theory of space constitute examples of the application of the second principle of the verifiability theory of meaning, i.e., of Ockham's principle; in modern physics, it was above all Einstein's theory of relativity which opened to Ockham's principle a new domain of application. It is not only the relativity of motion which we must mention here; there are also many other parts of Einstein's theories, such as his conception of simultaneity and his principle of equivalence of gravitation and acceleration, which are to be conceived as an outcome of the second principle of the verifiability theory of meaning. This principle may therefore be called the very basis of an antimetaphysical attitude.

What we said about the necessary expansion of the first principle of the verifiability theory of meaning is, however, valid for the second principle as well. Our insisting on the postulate of absolute verifiability would lead us to renounce any application of the principle because there are no sentences which can be absolutely verified. If we want to be able to point out sentences which have equal meaning, we must content ourselves with showing that they obtain an equal weight by all observable facts. We need not enter into a further investigation of this, as the discussion would only repeat the arguments of the analysis of the first principle.

As to the first principle, it was the effect of the transition from the postulate of absolute verification to the postulate of determinability of a weight that the domain of physical meaning was enlarged; propositions which had no meaning for the first conception obtained meaning for the second. Correspondingly, the same transition for the second principle implies an increase of the differences of meaning; propositions which have the same meaning within the physical truth theory of meaning may have different meaning within the probability theory of meaning. This occurs when the facts needed for the absolute verification of a proposition are not realizable for physical reasons, whereas there are facts physically possible which furnish different degrees of probability to the proposition in question. In our later investigations we shall discuss some examples of this kind (§ 14); they will show the importance which such a refinement of our logical instruments may obtain in the pursuit of the interpretation of the language of science and daily life.

If our expansion of the concept of meaning should be attacked on the ground that our wider concept of meaning might open the door to metaphysics, this would be entirely erroneous. Our theory of meaning is able to adopt Ockham's razor in a fitting form; the formulation we gave to the second principle cuts off all empty additions to sentences as well as does the formulation within the truth theory of meaning. The probability theory of meaning therefore maintains the antimetaphysical position of positivism and pragmatism, without taking over the too narrow conception of meaning from which these theories suffer if they are interpreted according to the strict wording of their programs.

Conversely, we must say that it is the probability theory of meaning alone which may give a satisfactory substantiation to the second principle of the verifiability theory of meaning. We pointed out that a substantiation of the verifiability theory of meaning consists in a relation between meaning and action; our example of the "divine animal" showed that we may co-ordinate to a given "super-empirical" proposition an empirical one which leads to the same actions. The second principle does nothing but formulate the consequence which this idea implies for a theory of meaning based on the relation of meaning to action. We may state it in the form: if two sentences will lead us under all possible conditions to the same actions, they have the same meaning. However, this formulation is possible only within the probability theory of meaning; for only if we introduce the predicate of weight can the relation of meaning and action be demonstrated. On the other hand, it becomes obvious from this formulation that the antimetaphysical function of the principle is kept. In our formulation also the principle denies any "super-empirical meaning" and states: *there is as much meaning in a proposition as can be utilized for action.* With this formulation, the close relation of the probability theory of meaning to pragmatism becomes still more obvious; we think, though, that our theory, by using the concepts of probability and weight, may furnish a better justification of the relation between meaning and action than pragmatism is able to give. This outcome of the probability theory of meaning—the connection of meaning and action—seems to me the best guaranty of its correspondence to empirical science and to the intention of language in actual life.

CHAPTER II

IMPRESSIONS AND THE EXTERNAL WORLD

CHAPTER II

IMPRESSIONS AND THE EXTERNAL WORLD

§ 9. The problem of absolute verifiability of observation propositions

The foregoing chapter was based on the assumption of the division of propositions into direct and indirect sentences. Direct sentences are sentences concerning immediately observable physical facts; such sentences—this was the presupposition—are absolutely verifiable, i.e., accessible to a determination of their truth-value within the frame of two-valued logic. Only for indirect sentences was the predicate of weight needed; such sentences are not controlled directly, but by means of their relation to direct sentences which confer on them a certain degree of probability.

This particular position of observation sentences, as direct sentences, is now to be examined. We must question their being accessible to direct verification. They deal with what is called a *physical fact;* our investigation, therefore, concerns the question whether we can verify a physical fact.

Before entering into detail, we must indicate that the word "fact" is used in a fluctuating sense. Sometimes physical laws are called facts because they are furnished by experience and not by deduction; but this is not what we shall here call a fact. Laws concern, on account of their claim to generality, an infinity of facts; we shall therefore distinguish them from facts, ascribing to this word a narrower sense.

To clarify our intention let us apply this distinction to some controversial examples. We know that the velocity of light is the upper limit for all velocities transmitting an effect; is this a fact or a law? According to our definition, generality characterizes law rather than fact, so this must be called a law. For the same reason we must call it a law that the Michelson interferometer shows the equality of the velocity of light in different directions because this result is stated for all apparatus of this kind. We obtain a fact if we proceed to consider the special experiment made by Michelson in 1883 with his special apparatus. To render the term more precise we may speak of a *single fact;* a single event, occurring at one definite spatiotemporal point, represents such a single fact.

We have now to apply our criticism to single facts and to ask whether single facts can be absolutely ascertained or whether propositions concerning single facts can be absolutely verified.

Let us consider the Michelson experiment. Every physicist knows that the statement concerning the equality of the velocity of light in different directions is not directly observed in the Michelson experiment but that it is inferred. Such a physical experiment is a rather complicated procedure. Directly observed are images in telescopes or on photographic plates, or indications of thermometers, galvanometers, etc. If we proceed from these experimental data to the statement concerning the velocity of light, this procedure is an inference, and an inference containing inductions. It contains, for instance, the presupposition that the temperature noted from time to time on the thermometer is valid also for the intervals of time between the moments of observation; that the laws of geometrical optics are valid for light passing through the telescope; that the lengths of the brass bars of the apparatus do not change during the observation (compared with other bars in rest relative to them), etc. It is obvious, then, that the statement concerning the velocity of light is not absolutely certain, being dependent on the validity of inductions. So this statement, although concerning a single case, is not absolutely verifiable. We see that mere reference to a single case is not sufficient to insure absolute verifiability to a statement.

We arrive at a more favorable result if we proceed from the statement concerning the velocity of light to statements concerning the individual data of the instruments used. It seems to be absolutely certain that at least the thermometer registered, say, 15° C. It might be a bad instrument, and the temperature of the room might be different from that indicated; but that this individual thermometer reached at this particular moment the line corresponding to 15° C.—is not this single fact absolutely certain?

This question leads us from the rather abstract facts of physics to the concrete facts of daily life. A thermometer is a thing built of glass, and mercury, and wood; a thing comparable to tables, chairs, houses, trees, stones—in short, a thing belonging to the sphere of our daily environment. To ascertain the existence of such objects requires no theoretical conclusions; so it seems possible to obtain absolute truth in this case at least.

It is well known that this assumption has been attacked by almost all philosophers since Descartes; and I should say for good reasons. The correct way of substantiating this attack seems to me to be the following one.

A statement concerning a physical fact, even if it concerns a simple fact of daily life, never refers to a single fact alone but always includes some predictions. If we say, "There was a table in my room, before my eyes, at 7:15

P.M.," this contains the prediction: "If no table passes the doors from 7:15 to 7:20, and no fire or earthquake acts on my apartment, then there will be a table in my room at 7:20." Or simpler still: "If I put a book on the table, it will not drop." It is because such predictions are included in the statement that it is not absolutely true, for an absolute reliability of the predictions cannot be warranted.

It might be proposed that we can separate these predictions from the statement, and reduce it to a bare factual statement; that is, that we exclude consequences concerning the table after five minutes, or concerning books placed on the table, and restrict the statement to the table just as it is seen. Such a reduction is possible; if we perform it, however, the statement loses its definite character. Saying, "There is a table," normally means that I maintain that what is referred to is a material thing capable of resisting the pressure of other physical things; this is what is expressed in the implication concerning the book. If I renounce implications of such a kind, the object I saw might be a picture furnished by a concave mirror; indeed everybody knows that illusions occur in which the image produced by a concave mirror is taken for a material object. The difference between the material object and the illusion cannot be otherwise formulated; it is only the consequences—i.e., future observations—which distinguish these two categories. This is the essential point. It might be objected that the future observations could be replaced by past observations—that I might have put the book on the table a moment before, or touched the table with my hand a moment before. But if I infer from this that the table as I see it now, without a book on it and without my touching it, is a material table and not the image produced by a mirror, then I perform an induction running, "If I were to touch it now, I would feel the resistance," or "If I were to put the book on the table now, it would not fall" —sentences which concern future observations and not past ones. It is true that past observations of the kind mentioned may suffice to substantiate my statement, but only because I base inductions on them; the statement concerning the table as a material object cannot be separated from predictions without losing its definite character; i.e., it would no longer indicate a definite physical object.

This, it seems to me, is the reasoning which proves indubitably that there is no statement concerning physical objects which is absolutely verifiable. Statements about simple physical objects are very sure but not absolutely sure. They are not sure because they are controllable; if we admit the possibility that later observation can control our statement about a present observation, we cannot exclude the case of a negative result of this control—that is, our statement cannot be maintained as certain. If in spite of that we take such statements as certain, we perform an idealization; we identify a high degree of probability with certainty. But, strictly speaking, this is not a case of truth but one of weight; even the observation sentences of daily life are not to be considered as direct sentences but as indirect sentences judged by the predicate of weight instead of the predicate of truth. The probability theory of meaning, therefore, is to be applied even to observation sentences of physics, or of daily life, if such sentences are to have meaning.

The attempt has been made to show that, although a physical statement never can be absolutely verified, it may at least be demonstrated in certain cases that the statement is false. If a book placed on a table does not stay lying there but falls down vertically, we might deem it sure that what is there observed is no material table. The principle of absolute verification, so we might suppose, might be replaced by a principle of absolute falsification.[1] Such an idea, however, is not tenable. Any falsification also presupposes certain inductions based on observations of other things and may be assumed with probability only. In our example it may be the book which is the nonmaterial thing, or which has become so the moment after withdrawing my hand from it; the statement about the material table then would remain true. Our statements about physical things are interwoven in such a way that the rejection of one of the statements may always be replaced by the rejection of another. Our choice as to the rejection is entirely made by reflections determined by the rules of probability. There is, therefore, no absolute falsification, just as there is no absolute verification. There remains nothing but the probability theory of meaning if we wish to justify observation propositions in the sense in which they are actually used in science or in daily life.

§ 10. Impressions and the problem of existence

The result of the foregoing section cannot be taken as a proof that there are no verifiable sentences at all. The uncertainty as pointed out concerns only observation sentences referring to physical objects. Philosophers who share our interpretation of sentences of such a kind have maintained the idea that there are observation sentences of another kind which can be absolutely verified. These are sentences concerning impressions. We must now consider this concept and ask for its epistemological significance.

The way in which so-called impressions are introduced is given by a continuation of the reasoning with which we questioned the truth of an observation sentence.

[1] This attempt has been made by K. Popper, *Logik der Forschung* (Berlin, 1935); cf. also my criticism of this book in *Erkenntnis*, V (1935), 267.

It is true that a sentence stating the existence of a material table implies predictions and that a reduction of it to a bare report would destroy its physical reference. But what, then, would be the result of such a reduction? We arrive, it is said, at a fact of another type: we come to say that at least I see a table. This is true whether it is a material table or the optical image of such a table produced by a concave mirror; so this at last is an indubitable fact. Facts of such a kind are called "impressions."[2] Thus there are, it is maintained, absolutely verifiable statements; what they concern, however, is not physical facts but impressions.

We shall accept, for the present, this conception. We shall admit that there are immediately given facts of such a kind, which the word "impression" or "sensation" is to denote—facts which we describe in sentences capable of absolute verification. A criticism of this assumption may be postponed to the following chapter. In the same way as the first chapter was based on the presupposition of the absolute verifiability of observation propositions, so the present chapter will be based on the presupposition of the absolute verifiability of impression propositions. It is only the consequences of this presupposition, not its validity, in itself, which we want to study for the present. We shall study these consequences by making use of the results of the foregoing chapter, which showed the relevance of the concept of probability; in a similar way, we shall show that the probability character of the inferences which occur affects the consequences resulting from the introduction of impressions as a basis of knowledge.

Impressions are—this is the usual conception—phenomena occurring within my mind but produced by physical

[2] The words "presentation," "sensation," and "sense data" are used in the same sense.

things outside my mind. Thus the concept of impression leads to the distinction between my own mind and the external world. Impressions are events of my personal sphere, of my private world; it is a grave mistake, so the adherents of this conception argue, to think that what I observe are things of an independent existence—I observe only the impressions produced by such things, i.e., effects of external things on my private world.

We said that we shall admit impression sentences as being absolutely certain; we see, however, that this absolute certainty is restricted to events of a private world only. With the transition from my own subjective experience to the objective external world, uncertainty enters into my statements. But not only uncertainty as to special statements; there is superimposed a general uncertainty as to the world of external things at all. How do we know that there is such an external world outside our private world? It is the problem of the existence of external things which arises here.

As long as we regarded observation sentences as the basis of knowledge, the problem of existence did not occur. There is no difference as to existential character between observational facts and other facts indirectly inferred; it is only the introduction of the basis of one's own psychical experience which creates the existence problem. This problem, therefore, is due to a certain advance in philosophical inquiry; it originates from the attempt to reduce knowledge to an absolutely certain basis.

Indeed, for the naïve conception of the world, there is no problem of existence. The sphere of daily life is not disturbed by the question whether the things we observe around us are real, are existent; any doubt of such reality would be considered ridiculous, as an outcome of an unhealthy departure from the clear views of daily experience.

The man of common sense is convinced that he is right in asserting that the tables, the houses, the trees, and the people around him exist as he does. Not only is this believed for objects of personal experience but the communications of other people and of scientific men are also accepted as certain. That there are other continents besides the one on which we live; that other planets and stars exist incomparably bigger than our small island within the universe; that unseen physical entities such as electricity, atoms, X-rays exist—all this is considered as a matter of fact which it would be simply unreasonable to doubt. This world of concretely existent things is further enriched by other things which are called "abstract," but which are nonetheless conceived as existent also. There is the state, as a political body, never directly seen as a whole, but whose reality is imposed upon everyone by daily experience; there is the spirit of the nation whose existence we find emphasized every day in the leading articles of the newspapers; there is the soul, our own and that of other persons, the doubt of which might lead to disagreeable collisions with the church; there is the financial crisis, the reality of which needs no confirmation by the holy authorities. In short, there is a solid and compact world around us, filled up by less solid but not less real things; this world is given to us from the early days of childhood, and there is no question as to its existence.

The beginning of doubt concerning this matter-of-fact world marks, indeed, a departure from the sound pursuit of daily affairs. It is that departure which leads from mere submission to traditional conceptions toward an intellectual penetration into the formation of concepts and marks the very beginning of philosophical thinking. It is the issue of the attempt to understand what we think, to clarify the bearing and the legitimacy of human conceptions. It is, therefore, an enterprise not less healthy than looking after everyday necessities; it is the sound desire to add to the struggle for existence an understanding of the struggle and of existence itself; and, if common sense attacks philosophy on account of its questioning fundamental concepts of life, this is only because the man of common sense does not realize that the desire for understanding may become as urgent as the desire for economic existence.

We preface this general remark to the following inquiry to meet the opinion of certain philosophers that an investigation of the question of the existence of external things is unreasonable and ridiculous. Such a position would be in itself an answer and would demand substantiation. It is true that the question of existence, as it is usually expressed, needs a correction; and it is precisely the task of the philosopher to clarify the question first before an answer can be given. But it is not legitimate to cut short the question by sophistical remarks. It has been argued by certain philosophers that a man who doubts the existence of external things ought to have his forehead knocked against a wall to convince him of the reality of the wall. I do not think this is philosophical reasoning. What the man saw might have better convinced him of external things than what he felt because what he saw was outside his body, whereas the pain he felt was inside; and it is just the question of whether there is something outside of himself which the man wanted to solve.

With this remark we are in the center of the problem of existence. Experience, even experience of daily life, compels us to distinguish between dreaming and being awake; there is a world of dreams as vivid as the waking-world—but nevertheless we know that we have to interpret this world as an internal world only, to which no external things correspond. Are we sure that the so-called "waking-world" is better? That this world is of a greater regularity is no convincing argument; nor is it an argument that in this world we even happen to reflect about its reality. That may happen in the dream world also; there are indeed dreams in which we try to discover whether we are in a dream and decide that we are not—only to discover on waking that this decision was itself part of a dream. The question concerning the reality of our waking-world, therefore, cannot be rejected as unreasonable; it is as reasonable as the distinction between the waking-world and the world of dreams.

§ 11. The existence of abstracta

There is a second problem of existence distinct from that of impressions. This is the problem of abstracta. What of the existence of such things as the political state, the spirit of the nation, the soul, the character of a person? Do things of such a kind exist? If they exist, are they things alongside of such concrete things as houses or trees? Or are they things of another sphere of existence? But what, then, is this other sphere? Since the time of Greek philosophy this question has been constantly discussed; it formed the subject of the famous controversy between nominalism and realism; it split philosophers into parties as thoroughly as did the question of the reality of the external world.

In spite of all differences there is one common feature in the structure of the two problems of existence. One questions the existence of abstracta as distinct from concreta, the other questions the existence of concreta in relation to impressions. It is this relational character which is common to both problems. We shall therefore have to study the relations occurring here. As these relations are of a simpler type in the problem of the existence of abstracta, we shall begin with this problem.

As to the problem of the existence of abstracta, it seems to me that the position of the realists was never a very good one. They insisted on the existence of abstract things, but they were always obliged to defend themselves by placing these things into a special sphere; the sphere of the "ideas" of Plato is the famous prototype of this kind of existence. There is, nevertheless, a strong natural feeling against such a procedure; the human mind needs a certain degree of perversion by sophistic training to be able to find some sense in such terms. The position of the nominalists, who maintained that only concrete things exist, looks much sounder, though I do not want to say that the ancient nominalists had already found the right form of solution.

The nominalistic idea is that abstracta are reducible to concreta, i.e., in terms of modern logic: that all propositions concerning abstracta can be translated into propositions concerning concreta only. To give an example: instead of saying, "The race of Negroes has its home in Africa," we can say, "All Negroes descend from forefathers who lived in Africa." In this way, the abstracta "race of Negroes" and "home" are eliminated and replaced by concreta, such as "descend" and "forefather"; the new terms which enter by this operation are logical concepts, such as "all." By the same method, such complex terms as the "political state" can be reduced to concreta. The logical method, in the general case, may be somewhat more complicated. It may turn out that to replace a statement containing an abstractum, more than one phrase containing concreta is needed. Thus the phrase, "The state is waging war," is to be translated into many propositions concerning soldiers, shooting, being wounded, and dying, men working in armament factories, others writing in offices, etc. We speak in general of a *reduction by co-ordination of propositions;* to one abstract proposition we co-ordinate a group of concrete propositions in such a way that the meaning of the group is the same as the meaning of the abstract proposition.

The equivalence of meaning on both sides of the co-ordination is an outcome of the theory of meaning as developed in chapter i. There is an equivalence of truth-value on both sides; if the abstract proposition is true, the group of concrete propositions is true, and if the abstract proposition is not true, not all concrete propositions taken in conjunction are true. It may be objected that in some cases the abstract propositions may be true even if not all concrete propositions are true; this may be because the same abstract fact may be realized by different concrete facts. The abstract fact, for instance, that there is good weather may be realized by a clear sky and a calm atmosphere, or by a partially cloud-covered sky and some fresh wind, etc. This case finds its logical expression by the introduction of disjunctions which allow us to maintain the equivalence in an expanded form. Let a be the abstract proposition and $c_1, c_2, \ldots,$ the concrete propositions; then the equivalence is to be formulated[3]

$$a \equiv [c_1 . c_2 \ldots c_m] \vee [c_{m+1} \ldots c_n] \vee \ldots \vee [c_{r+1} \ldots c_s] \quad (1)$$

In this way the exact logical construction of the abstracta is established. It follows from both the truth theory of meaning and the probability theory of meaning that both sides have the same meaning.

We see that the position of nominalism is connected with the verifiability theory of meaning; this, of course, is not a discovery of our time but the basic reason why both

[3] I use the signs of Russell: a period (.) for "and," $\vee$ for the inclusive "or," and $\equiv$ for logical equivalence.

theories have been developed in reciprocal relation. We have already mentioned that the nominalist Ockham was the father of our second principle of meaning. The nominalists were right in maintaining that the existence of abstracta is reducible to the existence of concreta.

What the ancient nominalists did not see was that it cannot be inferred from their theory of meaning that the abstracta do not exist. Whether or not we apply the category of existence to an abstractum is a matter of convention. We may say: "The race of Negroes exists." We know, then, that this means the same as, "Many Negroes exist, and they have certain biological qualities in common which distinguish them from other people." We may also say: "The race of Negroes does not exist." Then we have to add: "Many Negroes exist, and any proposition containing the term 'the race of Negroes' can be translated into propositions concerning those Negroes." We see, then, that the question whether or not abstracta exist, whether or not there is the term only or also a corresponding entity, is a pseudo-problem. The question is not a matter of truth-character but involves a decision—a decision concerning the use of the word "exist" in combination with terms of a higher logical order.

If we ask now which decisions are used in practice as far as the existence of abstracta is concerned, we meet the remarkable fact that there is no common rule, that the use of language decides sometimes for and sometimes against the existence of abstracta. To give some examples: the furniture belonging to a family is usually taken as existent; so is the company invited to a home, or a regiment of soldiers, or a court of justice. The decision is doubtful concerning such terms as "the state," or "human society," or "the third estate." In other cases there is a clear refusal to acknowledge existence: the height of a mountain does not exist, nor does the mortality of children, nor does left-handedness. The question of the motives of these decisions must be analyzed psychologically. It seems that those abstracta are conceived as existent with which we have concern in practical life, and which are usually expressed by nouns. We sometimes have to do with left-handed people, but we seldom employ the term "left-handedness"; so this remains a term without an existent object. Reference to "the furniture," however, appears frequently, and furniture is therefore conceived as an existent thing. The decision may even depend on the profession of the speaker. For a merchant, supply and demand may be existent entities, whereas an electrician would conceive an electrical charge as existent. It is a remarkable psychological fact that this "feeling of existence" which accompanies certain terms is fluctuating and depends on the influence of the milieu. The pursuit of this question is of great psychological interest; for logic there is no problem at all.

The possibility of ascribing existence to abstracta, however, does not justify the position of realism. The abstractum is not a thing of another "sphere" but a thing existing in the ordinary world. The furniture exists in the same world as the tables and chairs which form its elements; like these, the furniture is a thing which has weight and can be paid for in money. The realist introduces this other sphere because he believes in a surplus meaning of the abstract term. This is, I think, due to a misunderstanding of a logical fact which seems to have bothered ancient logicians, but which can be interpreted by nominalism without any difficulty. It is the fact that the abstract thing and the things which form its concrete elements cannot be "added," cannot be put alongside of one another. We are not allowed to count, say, a table and three chairs and a cupboard as six things, adding the furniture composed by these five things to them as a sixth thing. This is, however, a matter of the rules of language only; these rules contain prescriptions about the use of the terms "addition," "counting," "number," etc.—prescriptions which take account of the difference between the abstractum and its elements. To infer from this distinction the necessity of putting the abstracta into another "sphere" means mistaking a problem of language for a problem of being; a misunderstanding of the type which is responsible for the origin of the construction of so-called "ontology." The domain of the theory of abstracta has become a kind of maze composed of pseudo-problems.

Another pseudo-problem of this group is given by the problem of the spatial localization of certain abstracta. Does the state as a political body occupy a place in space? It may be answered that only the country belonging to the state, and not the state as a political institution, has a spatial extent. But this is a matter of convention only; it depends on the way in which we define spatial qualities. All qualities of the abstractum "state" are to be defined as relations between its concrete elements, so we may also define the spatial extent of the state as the space occupied by its inhabitants. The question whether a physical force is in space, or a melody, or the elasticity of a spring, is of the same type and is to be settled by a definition.

With these remarks the problem of the existence of abstracta finds its solution. This problem is a matter of decision and not a question of truth-character. Independently of the decision it may be stated that the existence of the abstracta is reducible to the existence of other things. This logical process may be called "reduction." The abstractum may be called a "complex"; the concreta on the right hand of formula (1) may be called the "internal elements" of the complex. The inverse process may be called "composition." The elements compose the complex, the complex is reduced to its elements. Both relations may be united into the term "reducibility relation"; it is defined by the equivalence (1).

Let us add a remark which concerns a relation with which we must deal in this context: this is the relation of the whole to its parts. This relation is to be considered as a special case of the relation of reducibility as defined. The parts are internal elements of the whole, as a complex. There is, however, no strict definition as to the use of this term. We use it when the complex has a spatial extent, and the elements have also spatial extents which form parts, in the geometrical sense, of the geometrical extent of the complex, as in the case of a wall and its bricks, or an estate and its grounds and fields. In this case the concept of the whole and its parts is reduced to the concept of geometrical whole and its parts. This conception is not always maintained, and sometimes the use of terms fluctuates; shall we consider the trees as parts of the wood? The definition of the relation of the whole and its parts is not given strictly enough to settle this question unambiguously. An example of a nonspatial case of this relation is given by a fortune and its parts, which may consist in cash, shares, and estates. It seems that we speak of a whole and its parts in a situation in which we ascribe to the elements certain numerical or geometrical quantities, the arithmetical sum of which is ascribed to the complex. This is, however, not a sufficient condition for the term. If the complex has, in addition, many other qualities which do not fulfil this relation, we do not consider it as a whole composed of its elements. The political state is usually not considered as a whole built up by its inhabitants as parts, though the quantity "total population" is the sum of the inhabitants; this is because the sum relation is not valid for many other qualities ascribed to the state.

Another example of the relation of reducibility is the case of the *Gestalt*. A melody is a *Gestalt* built up of tones; a drawing offers a *Gestalt* built up by pencil marks on the paper. This concept plays a great role in modern psychology, and for good reasons; but its logical nature as a special case of the relation of a complex to its internal elements has not always been pointed out by psychologists. They are right in saying that the *Gestalt* is not the "sum" of its elements, i.e., does not stand to these in the relation of the whole to its parts; but this does not imply that statements about the *Gestalt* have a surplus meaning over and above statements about its elements. On the contrary, the equivalence (1) holds here as well as in all other cases of the relation of reducibility. If this is disputed, the denial originates from an insufficient formulation of the statements about the elements, the relations between which must not be forgotten. The special conditions which a complex must fulfil to be called a *Gestalt* are, as yet, not so sharply demarcated that unambiguity is insured for all cases. This does not exclude, however, a useful application of the concept of *Gestalt* in many other cases.

The logical investigations which follow are independent of the special cases of the whole and its parts, or of the *Gestalt*. They concern the general case of the complex and its internal elements, expressed in the reducibility relation as formulated in (1).

§ 12. The positivistic construction of the world

We turn to the second problem of existence—the question of the existence of concreta. We begin our investigation with the consideration of the positivistic solution of the problem.

The positivistic conception of the existence problem may be summarized in one statement: The existence of concreta is to be reduced to the existence of impressions in the same way as the existence of abstracta is reduced to the existence of concreta.

This idea is an outcome of the positivistic conception of impressions as basic facts of knowledge (§ 10) in combination with the truth theory of meaning (§ 7). All observations are to be reduced, it is said, to impressions because it is only impressions that I can directly observe. Propositions concerning concrete physical things are, therefore, indirect sentences reducible to impression sentences as corresponding direct sentences; only the latter sentences can be directly verified. According to the principle of retrogression, this correspondence is an equivalence of meaning; therefore this correspondence is a reduction, in the sense defined in § 11.

Let us illustrate this by a simple example. The proposition, "There is a table," is inferred from certain impressions we have in looking at the table from different sides, by touching it, and the like. Now according to the principle of retrogression this inference is taken as an equivalence of meaning. Therefore the sentence, "The table exists," means the same as the sentence, "I have impressions of such and such kinds." It is the same relation as is valid for the reduction of abstracta; the table, therefore, is to be conceived as a complex, the elements of which are impressions.

This conception permits the positivists to interpret the existence of concreta in the same way as the existence of abstracta is interpreted. There is, they argue, no genuine problem of the existence of the external things; it is a pseudo-problem. We can say that external things exist; then this means the same as, "Impressions of such and such kinds exist." We can say also that external things do not exist. Then we must admit that the term "external things" may nevertheless be used and expresses the same as propositions concerning impressions. To decide upon the first or the second mode of speech is a matter of convention only. To demand more, to ask whether the external things exist "beyond" the impressions, would be meaningless. This is the famous positivistic interpretation of the existence of the external world.

It is one of the advantages of this conception that there remains no doubt as to the "reality" of the external world. The existence of the world is as sure as the existence of my impressions; this is because the first contention means no more than the second. Any doubt of the reality of the external world is an outcome of a meaningless question which supposes an existence of the things "beyond" my impressions. It would be the same meaningless question as to ask whether the race of Negroes has an existence of its own beyond the existence of the individual Negroes. To deny the existence of an external world, consequently, is not rejected as false but as meaningless; the positivistic solution, therefore, pretends to establish the world of external things in absolute certainty.

In spite of that conclusion, the positivistic conception need not deny a difference between dreaming and being awake. If we state a difference between the two, this must be inferred from a difference in impressions; this difference involves perhaps the great regularity of the impressions of the waking-state in comparison with the irregularity of the impressions of the dream. The whole of my impressions, therefore, may be divided into two classes such that alternately groups of impressions belonging to one class or the other follow one another; let us call these classes the "regular class" and the "irregular class." Applying the principle of retrogression, we find that the sentence, "I was dreaming," means, "My impressions belonged to the irregular class"; whereas the sentence, "I am awake," means, "My impressions belong to the regular class." The difference between dreaming and being awake is therefore saved by this theory; if anybody demands more, if he wants to maintain that the things he sees while being awake are "real" things whereas the things of the dream are "unreal" things, he says nothing because such a surplus contention would be meaningless. All that he wants to maintain by such words is sufficiently expressed by the already established difference between dreaming and being awake—because nothing else *can* be maintained.

These are the fundamental ideas of positivism as they are generally developed by their adherents. There is something very suggestive in these conceptions, something comparable to the convincing clarity of a religious conversion; and the ardor with which this interpretation of the existence problem has been emphasized by the preachers of positivism reminds one indeed of the fanaticism of a religious sect. I do not say this with the intention of discrediting positivism; on the contrary, it is just this strength of conviction which attracts our sympathies because of its manifest intensity and candor and its extreme desire to submit to the exigencies of intellectual cleanliness. But it is the danger of fanatic doctrines that they forget the necessary criticism of their basic conceptions; we must take care that admiration of the lucidity of the theory does not restrain us from a sober examination of its logical bases.

Our foregoing investigations of meaning lead us to an attack against one of the pillars of the positivistic doctrine. It is the principle of retrogression which we must question here. We found in § 7 that the relation between direct and indirect sentences is only a probability connection, not an equivalence. Thus the main idea of the positivistic reduction is not tenable. In the relation between abstracta and concreta, the co-ordination of propositions is an equivalence; only on account of this fact is the existence of abstracta reducible to the existence of concreta. If it turns out now that for the relation between concreta and impressions the co-ordination is of another character, the analogy does not hold; we are not then justified in saying that the existence of concreta is reducible to the existence of impressions. This means that the sentence, "The table exists," does not have the same meaning as the sentence, "I have impressions of this and this kind." The instinctive aversion we feel against submitting to the religious conversion turns out to have a sound logical basis. The positivistic interpretation of existence is not valid; there is a surplus meaning in the statement about the existence of external things. The positivist turns out to be a victim of the schematization which replaces a high probability by truth and takes the connections between propositions as relations ruled by the predicates of truth and falsehood. This schematization is allowable only for certain purposes; if it is made the basis for judging a question of principle, such as the question of the interpretation of existence, it leads to a profound discrepancy between epistemological construction and actual knowledge.

It will now be our task to develop another solution of the problem of existence—a solution in accordance with the probability character of the relations between propositions. To exhibit this solution we must first enter into a more detailed analysis of the nature of probability connections.

§ 13. Reduction and projection

We found that the transition from external things to impressions cannot be interpreted as a reduction; it is of another type of logical structure. To understand the nature of this structure, let us begin with the consideration of two examples.

The relation of reduction may be illustrated by the relation between a wall and the bricks of which it is built. Every proposition concerning the wall may be replaced by a proposition about the bricks. To say that the wall has a height of three meters reads in translation that there are bricks stuck together by mortar and piled upon one another to the height of three meters. The wall is a complex of the bricks; the bricks are the internal elements of the wall. The wall is not the "sum" of the bricks; this means that, if the bricks are separated from one another and scattered over the ground, the wall no longer exists, whereas the individual bricks may be unchanged. The wall is dependent upon a certain configuration of the bricks. This is included into our concept of "complex"; since all propositions concerning the complex are equivalent to propositions about the elements, the qualities of the complex will change if the relations between the elements change. The existence of the complex is dependent on certain relations between the elements, such that the complex may cease to exist whereas the elements still exist.

The inverse relation does not hold. If the elements cease to exist, the complex can no longer exist either. If the bricks are destroyed, the wall is also destroyed. This is what we mean by reducibility of existence: the existence of the complex is dependent on the existence of the elements in such a way that the nonexistence of the elements implies the nonexistence of the complex. This may be transformed into the statement that the existence of the complex implies the existence of the elements. The latter statement is only another formulation of the former. It is, however, to be distinguished from the converse relation according to which the nonexistence of the complex would imply the nonexistence of the elements, or the existence of the elements would imply the existence of the complex; as we saw, this inverse relation does not hold. There is, consequently, an asymmetry between the complex and its internal elements; it is just this asymmetry by which these two terms are distinguished, and which is meant by saying, "The existence of the complex is reduced to the existence of its internal elements." We do not make the inverse statement; the elements have, so to speak, a more solid existence.

It might be objected that a clever architect might be able to exchange the bricks, one after the other, for other bricks, in such a careful way that the existence of the wall remains undisturbed; the original bricks might even be ground to powder so that these elements no longer exist whereas the complex persists. This objection, however, is to be overcome by a more correct use of words. The wall made up of the exchanged bricks is a complex of other elements; if we speak nevertheless of the same wall, this complex "wall" is to be defined in such a way that it is constituted by one system, or another, of elements. That is, the complex is to be constituted by a disjunction of elements; or the propositions concerning the complex are equivalent to a disjunction of propositions about elements, as we stated formerly in the general formula (1) in § 11. Most of the complexes of usual language are of this complicated type. A melody may be played in different keys; the melody is defined by means of a disjunction of propositions. Our existence theorem, then, is to be formulated as follows: the existence of the complex implies the existence of *one* of the systems of elements but not the existence of a determinate one of the systems; and the nonexistence of *all* the systems of elements implies the nonexistence of the complex. We shall call such a complex a *disjunctive complex*.

We may give a more determinate form to the relation of the elements to the complex. We saw that the existence of the elements is not a sufficient condition for the existence of the complex. But it becomes a sufficient condition if some additional relations between the elements are fulfilled. If the bricks are arranged in such and such a way, the wall exists. Let us call these additional relations the *constitutive relations* between the elements. Then we may say for the simple as well as for the disjunctive complex: The complex exists if one of the corresponding systems of elements exists and fulfils the constitutive relations. This formulation expresses what we call the dependence of the complex on its elements. The elements may "produce" the complex; whether or not they produce it depends only on their internal relations. We must add, of course, that for this purpose the elements must be completely given; only in this case do we need no introduction of further elements to produce the complex. That is, only in this case can the constitutive relations be formulated with reference to these elements alone. Let us call such a set of elements a *complete set of elements*. The tones which the musician plays on the piano form such a complete set, that is, a set sufficient for the existence of the melody; it is not necessary to play other keys also. The constitutive conditions are formed here by the relations which constitute the temporal order of the tones, the length of the time intervals between them, and the like.

After this analysis of the concept of reduction we turn now to the consideration of another logical structure which is also characterized by a co-ordination of propositions, but which shows different qualities.

We imagine a number of birds flying within a certain domain of space. The sun rays falling down from above project a shadow-figure of every bird on the soil, which characterizes the horizontal position of the bird. To characterize the vertical position also, let us imagine a second system of light rays running horizontally and projecting the birds on a vertical plane which may be represented by a screen of the kind employed in the cinemas. We have, then, a pair of shadows corresponding to every bird; which of the shadows belongs to the same bird may be indicated by the outlines of the shadows. This correspondence allows us to determine the spatial position of every bird from the position of the corresponding pair of shadows and to determine the spatiotemporal movement of the birds from the spatiotemporal changes in the pairs of shadows. We can express this in the form of a co-ordination of propositions: every proposition concerning the movement of the birds is co-ordinated with a proposition about the changes of the pairs of shadows.

By this method the spatiotemporal position of the birds is projected into a system of marks which can be taken as a representative of the original birds. Analogous methods would allow us to construct marks for other qualities of the birds; for this we would have to employ other effects coming from the birds. The singing of the birds might be recorded, and the curved lines on the record would be the marks of the singing. Everything which can be observed from outside must be communicated to us by a physical process and can, therefore, be transformed into a physical thing outside the birds; this physical thing is our mark for the quality in question. We obtain in this way a system of marks which contains representatives for any quality of the birds observable from below, and which enables us to construe a co-ordination of propositions: every proposition concerning the birds is co-ordinated with a proposition, or a set of propositions, concerning the marks.

We contrive, in this way, to obtain a co-ordination analogous to the case of reduction illustrated in the example of the wall and the bricks. There are, however, some specific differences between the two cases; let us enumerate those qualities in which the second case differs from the first.

First, there is no equivalence of the co-ordinated propositions. This is because there is only a probability connection between the birds and the marks; if we see the marks only, we may infer with a certain probability that they are produced by birds, and if we see the birds only, we may infer with a certain probability that they will produce the marks. This lack of certainty is due to the fact that natural processes can never be foreseen with certainty. Whether or not the shadow-figures will be produced depends on numerous physical factors other than the presence of the birds alone, e.g., on the conditions on the screen. Conversely, whether or not there are birds as causes of observed shadow-figures cannot be inferred with certainty because there might be other physical processes having the same effect on the screen. Consequently there is no strict relation between the truth-values of the co-ordinated propositions. The proposition about the birds may be true, and that about the marks may be false; conversely, the proposition about the birds may be false, and that about the marks may be true.

Second, there is no reduction of existence. The birds have an existence independent of the existence of the marks. Using a mode of speech similar to our description of the existence qualities valid for reduction, we may say: neither does the existence of the birds imply the existence of the marks, nor does the existence of the marks imply the existence of the birds. The same is valid for nonexistence. This may be taken as a definition of what we mean by saying that the existence of the birds is not reducible to the existence of the marks. The shadow-figures may vanish while the birds still exist because other conditions may interfere; and the birds may be destroyed without the shadow-figures disappearing—because these may be produced by other physical causes.

In the example concerning the wall and the bricks we called the transition in question a *reduction;* in opposition to this we shall call the transition from the birds to the marks a *projection.* To express the parallelism we shall speak in both cases of a complex and its elements; to show the difference, however, we distinguish between a reducible complex and a projective complex, and call the elements of the former *internal elements*, the elements of the latter *external elements.* The birds are thus to be called a projective complex constructed by means of the marks as external elements. The most conspicuous feature of the projection is that it does not furnish a reduction of existence; this is because the relations between the projective complex and its elements are probability connections only. The probability character of these relations may be used to formulate the definition of the projection: A projection is a co-ordination of propositions, by means of a probability connection, in such a way that one term, or one set of terms, called the "complex," occurs only on one side of the co-ordination, and another term, or set of terms, called the "external elements," occurs only on the other side of the co-ordination. As the relation of probability connection is symmetrical (cf. § 7), there is no absolute difference between the elements and the complex of a projection; the terms may be interchanged. Thus the shadow-figures may be called a projective complex of the birds as external elements. Which side is denoted as the side of the elements depends on psychological conditions; usually we choose that side which is more easily accessible to observation.

To see the difference between both kinds of transition, let us consider a transition in which the birds are a reducible complex: this is the case when we consider as elements the cells of which the birds are constructed, or the atoms. These would be internal elements. An attempt might be made to conceive the projective complex as a disjunctive complex, by considering a disjunction of sets of elements which contains the internal elements as one set. But it is easily seen that the relations stated above for disjunctive complexes are not fulfilled. The existence of the complex implies, then, the existence of a determinate set of elements, i.e., of the set of internal elements; and it is not possible to add, to a set of external elements, constitutive conditions in such a way that the existence of the complex is implied. The projection is of a type logically different from a reduction.

Let us now apply the concepts which we have developed to the problem of the relation between impressions and external things. We pointed out that there is no equivalence between propositions concerning external things and propositions concerning impressions; there is only a probability connection. This relation is thus a projection and not a reduction; the existence of the external things is not reducible to the existence of impressions; the external things have an independent existence. It is the same kind of independence as between the birds and their shadows. Thus the naïve conception of independence of existence, as illustrated by this example, may be applied to the problem of external things and impressions as well; the idea that external things will persist after our death, when our impressions have vanished, may be conceived as valid in the same sense as the idea that the birds may persist when, on account of a cessation of the radiation, their shadows disappear. If we should consider, however, statements concerning external things as equivalent to statements about impressions, this would be interpreting the relation between external things and impressions as a reduction; so the existence of external things would be reduced to the existence of impressions. The external things, according to this theory, would vanish with the ceasing of our impressions—an idea which nobody seriously wants to maintain.

This interpretation of the existence problem will be attacked by positivism. We shall be answered that positivism does not maintain for external things and impressions a relation comparable to the relation between the wall and the bricks. Positivists agree with us in desiring to conceive the relation between external things and impressions as analogous to the relation between the birds and their shadows, i.e., as a projection. What they do not admit is that this relation of projection requires a probability connection. It is justifiable, they say, to talk of a projection also in a case when there are equivalence relations. What is to be altered for this purpose is only the form of the co-ordination of propositions. In the example of the wall the co-ordination is performed in such a way that the nonexistence of the bricks implies the nonexistence of the wall. There may be, however, another form of a co-ordination for which, in spite of the equivalence, the nonexistence of the elements does not imply the nonexistence of the complex. This can be attained if the existence of the complex at a certain time t_1 is defined by certain conditions valid for the elements at another time t_2. To give an example: We said that the melody is a reducible complex of the tones by which it is formed; we would substantiate this by saying that the melody vanishes when the tones disappear. We can, however, define the existence of the melody in such a way that the melody persists during the time intervals between the tones. We define: "The melody exists throughout the time-stretch running from the first tone to the last tone" means "There are tones at different individual times." Although the elements, the tones, do not exist in the time intervals between two tones, the melody does, and thus the existence relations for a projective complex are valid for the melody. This is even the usual way of conceiving the melody; for if we asked anybody whether the melody existed during all the time, from the beginning to the end of the music, he would surely answer in the affirmative.

To this objection we answer in the following way. It is true that such a definition of the complex may be given; but we are not obliged to do so—in the case of an equivalence we may always introduce another co-ordination for which the existence of the complex vanishes with the existence of the elements. The melody may be defined in such a way that it exists only at the moments when there are tones and vanishes in the intervals between the tones; such a kind of definition is equivalent to the one given above. Thus we arrive at an element of arbitrariness, just as has been already pointed out (§ 11) in the case of abstracta: the question whether or not the complex exists independently of its elements becomes a matter of convention. It is this arbitrariness which we do not accept for the problem of the existence of concreta. We maintain that a conception for which external things vanish with our impressions is not equivalent to the conception of an independent existence. Only in the case of probability connections is there no such equivalence; it is, therefore, only the conception of the projection as a probability connection between complex and elements which furnishes the admissible interpretation of the existence of the external world.

The preceding reflections nevertheless necessitate a slight correction of our interpretation of the reducibility of existence. We shall call the existence of the complex reducible to the existence of the elements when it is at least possible to introduce an equivalent system of propositions, in which the existence of the complex ceases with the existence of the elements. This definition of the term "reducible," however, does not require a change in our definition of reduction as a co-ordination for which all statements concerning the complex are equivalent to statements concerning the elements. The latter definition implies the possibility of defining the existence of the complex in such a way that the complex vanishes with its elements.

There remain some objections which we must now consider. They concern the question whether it is true that the probability connection can protect us from such consequences as pointed out for the equivalence connection, i.e., from the reducibility of the existence of external things to the existence of impressions. These objections will be considered in the following sections.

§ 14. A cubical world as a model of inferences to unobservable things

The objection which we consider first starts with questioning the analogy between the example of the birds and our situation in the recognition of external things. We said that the birds have an existence independent of their shadows on the screen; but to substantiate this we made use of the fact that there are other and direct observations of the birds which do not need any consideration of the shadows. We see the birds directly in their places within space; it is therefore easy to distinguish them from the shadows as different physical entities. In the case of our knowledge of the external world, however, we have nothing but impressions as a basis of the observation; is it logically possible to infer from here the separate existence of something which has an existence of its own, in the sense defined above, i.e., an existence which is not reducible to the existence of impressions?

This objection can be more precisely formulated in the following way. It is true that we use a probability inference when we infer from a given set of impressions to the existence of a physical thing. But is this more than an inference to new impressions? It seems impossible that by probability inferences the domain of impressions can ever be left; probability inferences, it may be supposed, will always remain within the domain from which they start. Thus statements about external things, in spite of the occurrence of probability inferences, will be equivalent to statements about impressions; not to statements about the observed set of impressions from which the probability inference starts but to statements about a certain wider set of impressions.

To discuss this objection it will be advisable to stay with the example of the birds at first and to carry through the discussion on this subject, since it is less exposed to misinterpretations. To obtain the same logical structure as in the problem of the inference from impressions to external things, we shall, however, alter this example in such a way that nothing but the shadows of the birds are visible. Thus we have comparable conditions in both problems.

We imagine a world in which the whole of mankind is imprisoned in a huge cube, the walls of which are made of sheets of white cloth, translucent as the screen of a cinema but not permeable by direct light rays. Outside this cube there live birds, the shadows of which are projected on the ceiling of the cube by the sun rays; on account of the translucent character of this screen, the shadow-figures of the birds can be seen by the men within the cube. The birds themselves cannot be seen, and their singing cannot be heard. To introduce the second set of shadow-figures on the vertical plane, we imagine a system of mirrors outside the cube which a friendly ghost has constructed in such a way that a second system of light rays running horizontally projects shadow-figures of the birds on one of the vertical walls of the cube (Fig. 2). As a genuine ghost this invisible friend of mankind does not betray anything of his construction, or of the world outside the cube, to the people within; he leaves them entirely to their own observations and waits to see whether they will discover the birds outside. He even constructs a system of repulsive forces so that any near approach toward the walls of the cube is impossible for men; any penetration through the walls, therefore, is excluded, and men are dependent on the observation of the shadows for all statements they make about the "external" world, the world outside the cube.

Will these men discover that there are things outside their cube different from the shadow-figures?

At first, I think, they will not. They observe black figures running on the screens quite irregularly, disappearing at the edges and reappearing. They will develop a cosmogony in which the world has the shape of a cube; outside the cube is nothing, but on the walls of the cube are dark spots running about.

After some time, however, I think there will come a Copernicus. He will direct telescopes to the walls and discover that the dark spots have the shape of animals;

and, what is more important still, that there are corresponding pairs of black dots, consisting of one dot on the

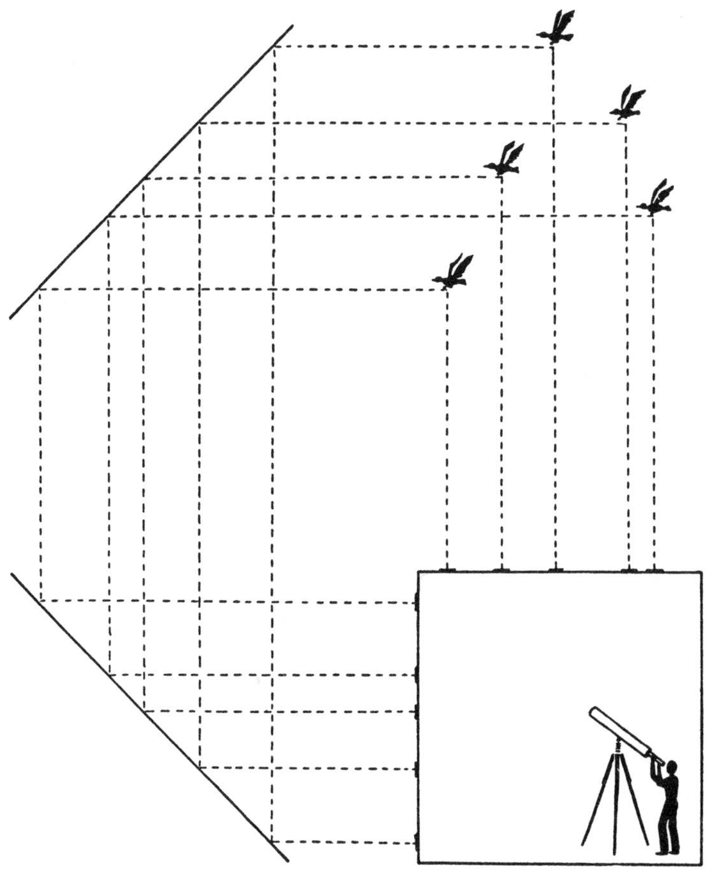

FIG. 2.—A cubical world where only the shadows of external things are visible.

ceiling and one dot on the side wall, which show a very similar shape. If a_1, a dot on the ceiling, is small and shows a short neck, there is a corresponding dot a_2 on the side wall which is also small and shows a short neck; if b_1 on the ceiling shows long legs (like a stork), then b_2 on the side wall shows on most occasions long legs also. It cannot be maintained that there is always a corresponding dot on the other screen but this is generally the case. If a new dot appears, whether or not there may be a corresponding dot already on the other screen, the new dot always starts from the edge of the screen but never appears immediately within the interior of the screen. There is no correspondence between the locomotions of the dots of one pair; but there is a correspondence as to internal motions. If the shade a_1 wags its tail, then the shade a_2 also wags its tail at the same moment. Sometimes there are fights among the shades; then, if a_1 is in a fight with b_1, a_2 is always simultaneously in a fight with b_2. It happens sometimes that one of the shades has its tail plucked out during a fight; then the corresponding shade on the other surface of the cube has its tail plucked out simultaneously. This is what is observed by means of the telescope.

Copernicus, after these discoveries, will surprise mankind by the exposition of a very suggestive theory. He will maintain that the strange correspondence between the two shades of one pair cannot be a matter of chance but that these two shades are nothing but effects caused by one individual thing situated outside the cube within free space. He calls these things "birds" and says that these are animals flying outside the cube, different from the shadow-figures, having an existence of their own, and that the black spots are nothing but shadows. I am, indeed, inclined to assert that such a Copernicus would arise among the people of the cube; the discovery of our real Copernicus, it seems to me, presupposed much more perspicacity and imagination.

The people, I think, would become convinced by this

theory; the question is, however, whether certain philosophers would be convinced. The positivists would attack Copernicus and argue in the following way:

What you maintain, they would say, is not false but biased. You say that there are things independent in their existence of the black dots; but you could say, on the same grounds, that these things are identical with the black dots. There is a correspondence between each of your "birds" and a pair of black dots; all that is said about your birds is inferred from the black dots and is therefore equivalent to statements about the dots. You believe in a surplus meaning of your hypothesis of the birds, compared with a description of the movement of the dots; but this is an illusion—both modes of speech have the same meaning. We admit your great discoveries concerning the relations between the dots, showing that there are corresponding dots on each of the two shade-covered surfaces of our cubical world. But your interpretation of this correspondence as an outcome of an individual identity of things outside the cubical world does not add a new content to your discoveries. This is only your way of speaking—other people prefer to speak of pairs of dots on the screens.

This means, in our terms, that the distinction between the projective complex and the reducible complex would be meaningless. Copernicus conceives the birds as a projective complex; the positivists answer him that he might conceive them, with equal reason, as a reducible complex with respect to the same elements, the black dots. The argument would be continued as follows:

We admit that this equivalence holds for our world only. If a man were once able to penetrate through the walls of the cube, he could distinguish between your hypothesis of the birds and the corresponding statement about the pairs of dots; if he were to see the birds above him, your hypothesis would be confirmed; if not, it would be refuted. But then there would be verifiable facts which distinguish your hypothesis from the pure description of the movement of the dots. For our world, however, there is a law of nature excluding any penetration of the walls of the cube; so, for our world, your hypothesis has the same meaning as the pure description of the dots.

In our terms, this argument would assert that the hypothesis of Copernicus has a surplus meaning only if we accept logical meaning, but that for physical meaning it has no surplus meaning when compared with the statement about the dots. It is this question which we now have to examine.

The positivistic interpretation is based on the presupposition of absolute verifiability. From within the cube, there is no possibility of obtaining a clear "yes" or "no" for the hypothesis of Copernicus; from an observation post outside the cube, such a clear distinction would be obtained. If we insist that only a clear "yes" or "no" is to be taken as an answer, the positivistic conclusion holds; this, I think, is the reason why the positivistic conception is so suggestive. It is, indeed, conclusive if we accept nothing but truth and falsehood as predicates of propositions; but it is no longer so if we introduce intermediate values—if we introduce the predicate of weight.

With regard to the predicate of weight the two conceptions are not equivalent. Judged from the facts observed the hypothesis of Copernicus appears highly probable. It seems highly improbable that the strange coincidences observed for one pair of dots are an effect of pure chance. It is, of course, not impossible that, when one shade has its shade-tail plucked off, at the same moment the same thing happens to another shade on another plane; it is not even impossible that the same coincidence is sometimes repeated. But it is improbable; and any physicist who sees this will not believe in a matter of chance but will look for a causal connection. Reflections like this would incline the physicists to believe in the hypothesis of Copernicus and to refuse the equivalence theory.

This means that the physicist insists on the surplus meaning of his interpretation not because it has logical meaning but because it has physical probability meaning. It is only physical truth meaning for which the positivistic interpretation is valid; but, if we admit physical probability meaning, there is a surplus meaning for the hypothesis of the birds (for the conception of the birds as a projective complex of the shades) because it obtains a weight different from that of the hypothesis of the pairs of dots, i.e., from the interpretation of the birds as a reducible complex of the shades. It is the different conception of the second principle of meaning which furnishes this distinction. The positivistic conception demands that two statements have the same meaning if they are equally determined as true or false by all possible facts; the probability conception demands the same meaning only if the statements obtain the same weight by all possible facts. It is to be admitted that the observable facts do not furnish a difference as to absolute truth or absolute falsehood of the two theories in question; but the weight conferred on them by the facts observable within the cube is different. Whereas the positivistic definition of meaning must therefore consider the two theories in question as having the same meaning, the probability definition of meaning furnishes a different meaning for both theories—although the domain of observable facts is the same, and although only the postulate of physical possibility is employed in the definition of meaning. The physicist, therefore, is not dependent on the acceptance of the dubitable concept of logical meaning and employs physical meaning as well as the positivist, but only in the probability form and not in the truth form.

The positivist, to defend his position, will answer in the following way: Your hypothesis, he will say to the physicists, obtains a different weight compared to my hypothesis only because it furnishes different consequences within the domain of our observable facts. Your theory, for instance, leads to the consequence that the coincidences between the shades of one pair will continue, will always be repeated; the conception that the coincidences are due to chance, however, leads to the contrary prophecy, to the consequence that the coincidences will not be repeated. To remove this difference we shall change our conception in such a way that it furnishes the same observable consequences as your hypothesis within the domain of observable facts, and that it differs only in the consequences for unobservable facts, for facts outside the cube. That is, we shall maintain our conception in such a way that the birds remain a reducible complex of the shades, but that all consequences for facts within the cube are the same as in the case of the birds being a projective complex of the shades.

This idea, if it were tenable, would prove that a difference between a reducible and a projective complex cannot be maintained, provided we keep to physical meaning.

Carrying through this idea the positivist would have to interpret the correspondence between the dots of one pair as a form of causal connection. He would have to say that there is a kind of coupling between the elements of one pair. If an element a_1 of one pair is approaching an element b_1 of another pair in a certain way called "fight," recognizable by a kind of excited dance of the shades and mutual bites with their beaks, there is—the positivist has to say—a causal effect transferred from a_1 to its corresponding dot a_2 on the other screen, and from b_1 to its corresponding dot b_2, in such a way that a_2 and b_2 enter into the same relations called "fight." With this hypothesis the positivist no longer would interpret the coincidences as chance but as an outcome of a causal law; and his conception, therefore, would furnish, as a consequence, the continuation of the coincidences for all the future. Thus his theory is altered in such a way that it does not differ from the physicist's conception as far as prophecies for future observable events are concerned.

The physicist, however, would not accept this improved theory. He is too clever a man to object to the positivist that such a causal connection is impossible; but he will say that it is very improbable. It is not because he wants to combine with the term "causal connection" some metaphysical feelings such as "influence from one thing to another" or "transsubstantiation of the cause into the effect." Our physicist is quite a modern man and needs no such anthropomorphisms. He simply states that, wherever he observed simultaneous changes in dark spots like these, there was a third body different from the spots; the changes happened, then, in the third body and were projected by light rays to the dark spots which he used to call shadow-figures. Freed from all associated representations his inference has this form: Whenever there were corresponding shadow-figures like the spots on the screen, there was in addition, a third body with independent existence; it is therefore highly probable that there is also such a third body in the case in question. It is this probability inference which furnishes a different weight for the projective complex and the reducible complex.

What is very remarkable here is that the two theories obtain, from the facts observed within the cube, different weights, although both theories furnish for future facts within the cube the same weights.[4] The probability conception of meaning, therefore, allows us to distinguish between theories which furnish, for all observable consequences of a certain domain, the same weight, even if nothing but facts of this domain are at our disposal for the probability inferences.

It may be said that this is possible only if the theories in question differ at least in logical meaning. This is not false; as we have already pointed out, two theories which have the same logical meaning cannot obtain different probability meaning. But the concept of probability meaning has the smaller extension; not all propositions having different logical meaning have also different probability meaning. We cannot say, therefore, that we accept the theory of the physicist as meaningful because it has logical meaning. We accept it because it has physical probability meaning.

We might attempt another substantiation of the necessity of accepting logical meaning. It might be said that, although not every difference of logical meaning renders a difference of probability meaning, those cases in which the difference occurs can be carried through only on account of the difference in logical meaning. To speak more clearly: if we could not at least imagine a difference in logical meaning, it would not be possible to calculate a different weight for the two theories. But this, I think, would be a grave mistake. The concept of logical meaning is valid

[4] Remark for the mathematician: There is a relation between the "forward probabilities" from the theory to the facts and the "backward probabilities" from the facts to the theory; this relation is expressed by the rule of Bayes. But in this rule there occurs still a third set of probabilities usually called misleadingly "a priori probabilities," or, better, "initial probabilities." It is these initial probabilities which are involved in the reflections of the physicist about causal connections. Thus the "backward probabilities" may be different, although the "forward probabilities" are equal, on account of different "initial probabilities."

only within the sphere of idealization in which physical propositions are taken as absolutely verifiable; if we take into consideration that truth signifies, strictly speaking, nothing but a high weight, we find inversely that truth meaning is to be reduced to probability meaning. We see this if we consider once more our example of the birds. The objection here would read thus: You are entitled to infer, with probability, that there are birds outside the cube only because you can at least imagine that you penetrate through the ceiling and see the birds; this penetration, although excluded by a law of nature, is logically possible, and therefore the object of your probability inference has meaning. The fault of this reasoning becomes obvious if we now introduce the case of a penetration of the ceiling. If a man were able to pierce a hole through the ceiling, and to see the birds—would this be an absolute verification of the theory of the cube-Copernicus? We have shown that there are no statements capable of absolute verification. The man could construct an interpretation for which the birds were not material bodies but only optical images produced by light rays coming from the shadows, deflected in such a way that the rays coming from the dots of one pair met at a certain point in space and ran from there into the observer's eyes. Relative to what one sees this cannot be called false but only very improbable. So what is obtained by a "direct observation" is an increase of weight for the theory of the birds but not a verification. The objection in question, therefore, would finally maintain that a theory can be meaningfully inferred with probability only if it is at least logically possible to construct facts which confer a higher degree of probability on the theory. I do not think this conception will be seriously maintained.

Statements made in terms of the later verification of a theory which is for the time being only rather probable on the basis of observed facts have the advantage of being an intuitive representation of the theory—but they are not the sole form in which the meaning of the theory is to be expressed. To say, "The statement that the birds are a projective complex of the shadow figures means that, if we should penetrate through the ceiling, we should see the birds," is only a short and intuitive way of expressing what is meant—nothing more. In this way we pick out one of the consequences of the theory which if observed would make the theory highly probable; but by no means do we obtain by this method the full meaning of the theory. What we get is an intuitive representation of the theory. We say, for instance: " 'Next year there will be a European war' means 'There will be airplanes above London, and shooting, and wounded men in the hospitals.' " Or we say: "A visit to New York means seeing skyscrapers and streets full of cars and men rushing for business." In this way we take certain representations for the whole; but it must not be forgotten that many other features are dropped by this method. The method is the more dangerous in case the chosen representations are not physically accessible but only accessible to our imagination. This is the case when it is physically impossible to obtain high degrees of weight for a theory. It may be advantageous, for certain purposes, to visualize the statement by imagining just the inaccessible results which would furnish the higher weight; but it must not be forgotten that we then obtain a representation only. Thus it may be permissible to visualize the concept "atom" by imagining the impressions of an observer whose size is of submicroscopic dimensions. But to insist in such cases that only the facts conferring a high weight on the theory are to be taken as its meaning is an outcome of the schematized conception of the two-valued logic. Actually, such a division of facts does not correspond to the practice of science. Considering observations of the physically inaccessible domain, we do not obtain facts which verify statements concerning things situated there but only facts which confer a higher weight to such statements. But then there is only a difference of degree with respect to statements based on facts observed within the accessible domain. The probability theory of meaning, therefore, does right to admit statements as different in meaning if these statements obtain different weights from observed facts—without regard to the question whether or not there will be, later on, a better determination of the weight.

It is, however, not false to employ the concept of logical meaning in the sense of a meaning defined by the logical possibility of obtaining a high weight. We may say that physical probability meaning is a domain between physical truth meaning and logical meaning; it allows us to make inferences which infringe upon the domain of logical meaning, although it is based on the physical possibility of ascribing a weight. The probability theory of meaning therefore allows us to maintain propositions as meaningful which concern facts outside the domain of the immediately verifiable facts; it allows us to pass beyond the domain of the given facts. This *overreaching* character of probability inferences is the basic method of the knowledge of nature.

An example taken from physics may illustrate the significance of the probability theory of meaning. Einstein's theory of relativity forms the famous domain for examples of the application of the verifiability theory of meaning; but, if we consider this theory more exactly, we find that it is physical probability meaning, and not physical truth meaning, which is here applied. Let us consider Einstein's theory of simultaneity. We send at the moment t_1, from the spatial point A, a light signal to the spatial point B, arriving there at the moment t'_2; here the signal is reflected and returns to A at the moment t_3. t_2 may be a moment at A, between t_1 and t_3, but arbitrarily chosen in this interval. Then, according to Einstein, the statement s, "t_2 is absolutely simultaneous with

t'_2," has no meaning. This is usually substantiated by saying that this statement is not verifiable, i.e., has no physical truth meaning. This is, however, not correct; Einstein maintains more—he maintains that the statement s cannot be provided with a weight, and so has no physical probability meaning. Just because probability meaning is a "more tolerant" concept than physical truth meaning, the denial of probability meaning is a stronger postulate than the denial of physical truth meaning.

To show this, let us first note that the statement s has logical meaning. This reads: "If there were no upper limit to the velocity of signals, a signal of infinite velocity[5] leaving B at t'_2 would reach A at t_2." This, of course, would be true only for one determinate t_2 between t_1 and t_3, so that this time-point is distinguished as absolutely simultaneous to t'_2. For any other t_2, the statement would be false; but then it has meaning as well. We are allowed, therefore, to say that the statement s has logical meaning for every t_2. If Einstein rejects the statement s, he decides in favor of physical meaning. But he demands more than physical truth meaning; he demands that all other facts of nature are of such a kind that they do not furnish, for a determinate t_2, a higher probability of being a specific time-point than for other values of t_2.

Such a distinction might be given by the transportation of watches. Einstein demands that two watches equally regulated during a common stay at A, and moved in different ways and with different velocities toward B, will show at B, after their arrival, a difference in their readings. We can imagine a world in which this is not the case, but in which the indications of two watches are in correspondence after the different transportations from A to B. In this world transported watches would define a simultaneity which we call *transport time*,[6] and we would say: If there were no upper limit to the velocity of signals, the infinite velocity would determine with a great probability, as simultaneous to t'_2, *that* time-point t_2 which corresponds to the transport time. In this world absolute simultaneity would have a physical probability meaning, though no physical truth meaning. Einstein refuses to believe in the existence of experiments, like the described transportation of watches, which would distinguish a certain t_2 as probably being the time-point of the arrival of infinitely quick signals. Thus Einstein refuses physical probability meaning to absolute simultaneity, which is, as we see, a stronger postulate than the refusal of physical truth meaning.

[5] The concept of infinite velocity may here be eliminated and replaced by a more complicated statement about the limit of the times of arrival belonging to signals of finite velocity, which defines a fictive "first signal" (cf. the author's *Axiomatik der relativistischen Raum-Zeit-Lehre* [Braunschweig, 1924], p. 24).

[6] *Ibid.*, p. 76.

Our conception of the example of the cubical world, which accepts the statement about the birds outside the screens as meaningful and different from statements about the dots on the screen, is therefore not in contradiction to the principles of modern physics. The cubical world as described would correspond not to Einstein's world but to a world in which a transport time would be definable. The principles of the theory of relativity have been wrongly interpreted as suppoıts for the concept of physical truth meaning; what they actually support is the concept of physical probability meaning.

§ 15. Projection as the relation between physical things and impressions

We proceed now to the application of our concepts of reduction and projection to the problem of the existence of the external world.

By analogy with the example of the cubical world our contention reads: Impressions are only effects produced within our body by physical things, in the same sense as the shadows are effects of the birds. Thus impressions are only external elements relative to the physical things; these things are projected to our impressions but not reduced to our impressions. The "external world" therefore has an existence of its own, independent of our impressions.

This is the so-called realistic conception of the world. Let us see what positivism answers. The answer is known to us from the example of the cubical world. It reads:

"What you maintain is not false but biased. You say there are things independent in their existence of your impressions; but you could say, on the same grounds, that these things are a reducible complex of your impressions. There is a correspondence between your impressions and your external things; all that is said about your external things is inferred from impressions and is therefore equivalent to statements about impressions. You believe in a surplus meaning of your hypothesis of the external world;

but this is an illusion—both modes of speech have the same meaning."

We need not repeat our discussion of this objection. We summarize only: It is not true that our statements concerning external things are equivalent to statements about impressions, although they are inferred from them. It is not true that the statement, "The external world is a reducible complex of impressions," has the same meaning as the statement, "The external world is a projective complex of impressions." This might be said, perhaps, if we accept physical truth meaning; but then there are no physical statements at all because there are no absolutely verifiable statements about the physical world. If we want to obtain meaningful statements, we must introduce physical probability meaning; and then the assumed equivalence between the reducible complex and the projective complex does not hold. There is a surplus meaning in saying that there is an external world independent of our impressions.

The reason, it seems, why positivists maintain this equivalence is to be found in their idea that it is not possible to infer from a certain domain of things to another domain. It is the neglect of the overreaching character of the probability inference which leads positivists to their equivalence theory. They believe that we are obliged to interpret probability inferences by the principle of retrogression, and so they do not see that the probability inference passes beyond the given observations. This error about the logical nature of the probability inference is the root of the positivistic doctrine of existence.

To clarify this error, let us consider the application of the principle of retrogression to probability inferences. Thus we come back to a form of the positivistic argument stated in the beginning of § 14. Let *i* be the conjunction of statements about the impressions (forming the class I) from which the probability inference starts and e the statement about external things (forming the class E) which is inferred from i with probability. It is true, then, that i is not equivalent to e. But what is maintained is that there is a more comprehensive conjunction i' of statements about impressions (class I'), including predictions about future impressions, which is equivalent to e.

Let us ask whether there is such a conjunction i'. The first thing we can say is that if there is such a class the corresponding class I' cannot be finite, as the observable consequences of a physical statement do not form a closed class.[7] But we can say more. Even statements about an infinite class of impressions are not equivalent to the physical statement. This becomes obvious if we consider impressions as physical effects caused in our body by the external object and apply a general theorem concerning causes and effects.

If we have a cause and collect from all its effects a certain class which may be infinite, but which does not contain the cause itself, the cause and the class of effects stand in the relation of projection; a statement about the cause is not equivalent to any set of statements about the class of effects. They are in a probability connection only. The statement, "The sun is a ball of glowing gases of high temperature," is not equivalent to any set of statements about physical facts outside the sun, even if the set is infinite and even if it comprehends all points of a surface surrounding the sun; we get by these observations a set of elements from which we may with probability infer the sun's exist-

[7] We have to take account of the fact that an infinite class of impressions may be described by a finite class of propositions. If we say, e.g., "If there is a gravitational field at all points within a certain space, the impression of heaviness is obtainable"; this is one proposition, but it concerns an infinity of impressions. The denial of this sentence would also require an infinity of observations.

ence and qualities, but which is by no means of equivalent meaning. Only if we were to include the sun itself in the set of observed facts, would there be an equivalence; but in this case all other facts might be dropped, and nothing would remain but a trivial tautology.

There is no difference if the effects produced consist of impressions. We cannot say, therefore, that there is a conjunction of statements i' to which e is equivalent. This would be permissible only if I' were to include the physical object, i.e., if we include the case that our body might become identical with the physical object. This is not logically impossible; but the positivist will scarcely be ready to accept this idea as the only correct interpretation of his thesis that there are statements about a class of impressions which are equivalent to the physical statement. This would mean that a statement about the sun is equivalent to a statement about impressions because it is not logically impossible that one day the sun may be a part of my body, and the movement of its glowing gases signifies, within myself, an observational process. We may leave this interpretation to the novelist, I think, and keep to our probability theory of meaning which needs no such equivalences.

We have to say, therefore, that the physical statement e is not equivalent to statements i' about a class I' of physically attainable impressions. We cannot determine a class I' of impressions such that, if i' is true, e is also necessarily true. This is what I call the overreaching character of probability inferences in application to the problem of impressions and the external world. The nonequivalence between e and any conjunction of statements i is what is meant by saying, "The external things have an existence of their own independent of my impressions."

To show the failure of the positivistic equivalence theory, let us consider an example. We take the proposition, "External things will continue to exist when I am dead." Common sense is convinced that this proposition, if it is true, may be considered as a proof that the existence of external things is not reducible to the existence of impressions; external things are, on the contrary, to be conceived as a projective complex of impressions. The positivist maintains that both interpretations are equivalent; so he has to say that the proposition, "External things will cease to exist when I am dead," has the same meaning as the former. Let us give to both propositions a more precise formulation. The first, which may be called e_1, is to read: "Until and after my death, external things will persist as is usually expected." The second proposition e_2 may be: "Until my death external things will persist as is usually expected; but, after my death, external things will vanish." If the positivist maintains that these two propositions e_1 and e_2 are equivalent, the reason lies in the fact that both hypotheses have the same observable consequences, or, strictly speaking, that they furnish the same weight for all possible predictions which I can make for the stretch of life lying before me. But we saw, nevertheless, that such hypotheses may obtain different weights from the observable facts. This is obviously the case here. Seeing that many people who are similar to me expire without producing such fatal consequences to the physical world, I infer with high probability that the same will be the case when I die. This is a correct reasoning comparable to a great number of similar inferences occurring in physics and never questioned there because they do not concern my own person. Thus the probability theory of meaning furnishes a different meaning to both sentences and accords with common sense.

Introducing the concept of logical meaning, we could

also say that the proposition e_1 is meaningful and different from e_2 because it is logically possible that I awake, after my death, and verify the existence of the physical world. This interpretation is permissible in the sense stated above, as an intuitive representation of what is meant. But, if we were to accept this interpretation as the only justification for statements about the world after our death, we would be led into great difficulties. As we have pointed out (§§ 6, 8, 14), logical meaning is too wide a concept; it is not compatible with the conceptions of modern physics. Thus a man who accepts a statement about the world after his death as meaningful only because it has logical meaning would be obliged to accept absolute simultaneity as well. On the other hand, a relativist who insists on the postulate of absolute verifiability would be obliged to consider statements about the world after his death as meaningless. It is only probability meaning which leads us out of this dilemma, justifying jointly the statement about the world after my death, and the rejection of absolute space-time conceptions.

It is not always an easy matter to discuss this question with positivists. They usually become offended when they are told that they do not believe in the existence of the physical world after death. They emphasize that this is a misunderstanding of their theories and demonstrate their conviction of the persistence of the external world after their death by taking out life insurance policies in favor of their families. They do not acknowledge our reasoning but insist that for them also there is a difference between the statements, "The external world persists after my death" and "The external world does not persist after my death." The difference is, they say, that the first statement includes certain statements concerning the death of other people without the world's being annihilated, whereas the second statement would contain statements about the world's vanishing simultaneously with the death of other people. This, however, is not the problem in question. The two statements we previously formulated are not the same as the two statements compared by the positivist. The second statement, in our formulation, reads otherwise. We formulated it in such a way that the difference of the two statements begins only with my death, saying that until my death all should be the same as usual. These statements cannot be distinguished within the positivistic theory of meaning, i.e., by means of the concept of physical truth meaning. I do not doubt the seriousness of the positivists as far as life insurance policies are concerned; what I want to maintain is that they cannot justify this carefulness because their theory furnishes no means of distinguishing between the statements e_1 and e_2 formulated by us.

§ 16. An egocentric language

We showed in the preceding section that propositions about external things are not equivalent to propositions about impressions. To give a new illustration to this conclusion, let us now consider an objection which attacks our result from another point of view. This objection starts from reflections which we introduced at the end of § 13. We showed there that in the case of a reduction the relation between the complex and its elements may be defined in different ways. Only for one kind of co-ordination of propositions does the existence of the complex vanish with the existence of the elements; for another kind of co-ordination, this consequence may be avoided. We maintained that the possibility of a co-ordination which has this consequence will suffice for us to call this case a reduction, and the complex a reducible complex. It may be objected, however, that perhaps the situation in the case of probability connections is not otherwise; that in this case it is also possible to introduce a co-ordination of propositions for which the existence of the complex vanishes with the existence of the elements. If this is true, it will show that there is no genuine difference between projection and reduction, but that this is a difference of language only. The objection in question, therefore, is proved as valid if we succeed in constructing a language for which the existence of the projective complex is dependent on the existence of the elements.

We shall find a way to construct such a language by starting from the very contention which we intend to actualize in our new language. We shall try to exclude the independent existence of external things by establishing this idea in the form of a principle which we make the basis of our language. To facilitate our task, let us consider an example. Let us imagine a man who is convinced that all things cease to exist as soon as he ceases to look at them; how could he defend his conviction against the objections made to him by common sense and by scientific thinking? He could defend himself if he had sufficient imagination to invent complicated logical constructions which connect the different impressions perceived by him in certain time intervals. He could interpret the reappearance of the things at the moment when he looks at them by saying that his looking produces the things. Thus he has to introduce a new kind of causality; but, if he is careful and consistent, he can carry through his conception. There are experiences which show that there is a certain "development" in a physical state. We put a kettle of cold water on the fire, come back after five minutes, and see the water boiling. The man in question would have to say that his looking at the kettle produces the things in the same advanced state which the things would have acquired by their intrinsic development if he had observed them and had not interrupted their existence. His new causality thus obtains strange qualities but not impossible ones. He will find even stranger qualities when he takes into consideration observed effects produced by the things at a moment when he does not look at the things. He looks at a tree and observes it as existent; then he turns and no longer sees the tree but its shadow. His conception, then, compels him to say that there is an aftereffect of the tree—the shadow—which persists for a long time when the tree itself has already vanished. This would mean a change in the laws of optics; but it could be consistently carried through.

Would this conception ever lead to contradictions with observable facts? Obviously not, because all experiences are interpreted by the same principle. The laws of optics as obtained by this man from experience would differ from our laws of optics. They would be divided into two classes by means of the clauses "if I observe certain things" and "if I do not observe certain things." The laws of the first class are equal to our laws; the laws of the second class, however, speak of strange aftereffects and things appearing fitfully in different states of development. This furnishes a rather complicated description of the world, but it does not lead to any contradiction of experience. If there is a seeming contradiction, this is only because the distinction of the two classes of phenomena has not been consistently carried through; it can, therefore, be eliminated by a change of interpretation.

We may raise the question whether the hypothesis of this man, though at least compatible with the facts, does not obtain a rather low degree of weight, i.e., can be demonstrated as being very improbable. It turns out that even in this respect there is no difficulty for him. There is one kind of experience which might be considered as a difficulty: the man sees that, when other persons turn their eyes away from things, these things still persist. If he admits the similarity between himself and other persons, this would render a high probability that the things will also persist when *he* does not look at them. But this is only valid under the presupposition of the similarity mentioned; so our hypothetical man may turn his inference in the opposite direction and maintain: I have an exceptional position in the world because the things vanish only when I do not look at them, whereas they persist when other persons do not look at them. When this conception is introduced, the probability inference from other people's nondisturbance of the existence of things to his nondisturbance of them is not valid. The methods of probability, therefore, do not furnish a result which throws into question the hypothesis of our example.

We may be astonished at such a result. We have so far maintained that the existence of things which are not observed may be inferred with high probability, even in the case when a direct observation of the things is excluded by certain physical laws, as in the case of the birds and the cubical world. We find now that we can introduce another conception for which the things do not exist at all when they are not observed and that this conception may obtain a high degree of probability as well. Is not this a contradiction?

The seeming contradiction is dissolved when we enter into a more detailed analysis of the second conception. We shall find, then, that our plan of constructing another language has been actualized in our example—that the man who conceived the things as vanishing when he does not observe them speaks another language than we do and that the apparent contradiction is due to a different meaning of words. This is to be understood in the following way.

Any description of the world presupposes certain postulates[8] concerning the rules of the language used in the description. The description of unobserved facts depends on certain assumptions concerning causality and therefore depends on postulates about causality. The postulate normally in use for this purpose requires us to construct homogeneous causal laws, as far as it is possible. The last clause is necessary because it is not always possible to construct homogeneous causal laws; thus it is not possible to construct for things seen in a dream the same laws as for things seen during waking. (Things seen in a dream are not seen once more in the next dream, etc.) But experience shows that for the things seen in the waking-state it is possible to describe the state of things during the interval between two observations in such a way that the principle of homogeneity of causality is satisfied.[9] This is done when we consider the things as existent during these intervals, whereas considering the things as nonexistent implies changes of causal laws, as we found in our example. The postulate of the homogeneity of causality, therefore, decides in favor of the conception of the existence of nonobserved things.

The man who conceived nonobserved things as nonexistent, however, decides in favor of another postulate. He

[8] Whether or not these postulates are conventions must be specially examined (cf. the remarks about equivalent and nonequivalent languages in § 17).

[9] There is, strictly speaking, a difference between homogeneity of causal processes and homogeneity of causal laws. The first postulate demands that the causal processes in physical things are not disturbed by our observation; the latter postulate demands only that, if there is a disturbance, this is to be according to causal laws for other phenomena. The first postulate cannot always be maintained; we know that scientific instruments of a more sensitive type are disturbed by the observer (by slight mechanical impacts, by the change of temperature caused by the observer, etc.). Quantum mechanics has even shown that there is a principle of disturbance by observation which cannot be reduced below a certain minimum. The second postulate, the equality of causal laws for the disturbance by the observer and for other physical phenomena, has turned out to be always maintainable in modern physics.

renounces a postulate concerning causality; his alternative postulate is the principle that things do not exist when they are not observed. Thus this assumption is for him no empirical matter but a decision and, therefore, beyond question. His scientific language, however, is altered by this procedure, and we must now point out in what respect.

The first and very obvious change is that his word "existence" does not correspond to our word "existence" but to our word "existence observed by me," or, simply, "being observed by me." Let us call the language of the man the *egocentric language;* then we may establish the following correspondence:

Egocentric Language	Usual Language
1. Things do not exist when I do not observe them.	1. Things are not observed by me when I do not observe them.
2. Things are produced at any time when I turn my eyes in a certain direction.	2. Things are observed by me at any time when I turn my eyes in a certain direction.

Both propositions are not about things directly but about observations of things. The first is a tautology, as is obvious in the expression within our usual language; this is because this proposition is nothing but the formulation of the postulate accepted by the man in question. The second is not always true, as it may happen (expressed in usual language) that the thing has been removed, or disappeared, while I was turning away; this character of not being always true is valid for both languages.

Let us now try to express a sentence which concerns not our observation of the thing but the independent existence of the thing. Take the sentence, "The thing exists during a certain time interval Δt." We pronounce such a sentence if we observe the thing at least at certain moments within the interval Δt, or if we discover that the observation is prevented by other things which do not exclude, however, that certain effects of the thing are observed by us. A stone which we saw may be covered at a second observation by a person, whereas the shadow of the stone is still to be seen. We express this idea in the following way in both languages:

Egocentric Language	Usual Language
3. If I turn my eyes during the time interval Δt in a certain direction, the thing is produced, or I can construct, applying my causal laws to the things which I observe, a cause which prevents the thing's existence. This cause must be of such a kind that it does not prevent the existence of certain other things which would be, if the thing were to exist, according to my causal laws the effect of the thing.	3*a*. If I turn my eyes during the time interval Δt in a certain direction, the thing is observed, or I can construct, applying my causal laws to the facts which I observe, a cause which prevents the thing's observation. This cause must be of such a kind that it does not prevent the observation of certain other effects which my causal laws ascribe to the thing.

This sentence is of a better truth-value than sentence 2 because it takes into account the possibility of a disturbance of the observation. But even sentence 3 is not always true; it may happen that the thing (in usual language) really vanishes, as a cloud may vanish by being evaporated. Thus sentence 3 can be pronounced only with a certain probability, although with a higher probability than sentence 2.

The question remains: Is proposition 3 equivalent to the usual proposition, "The thing exists during a certain time interval Δt"? This means: Do we call the latter proposition true, when proposition 3 is true, and inversely? It is obvious that this is not the case. We can only say that, if proposition 3 is verified, there is a high probability for the thing's existence; and conversely, if the thing exists, there is a high probability for proposition 3. The first statement is an expression of our general idea that observations never can furnish an absolute verification of a sentence about physical things. The second statement takes into account the case of an exception to the known rules of causality; it might happen that the laws of optics are suddenly superseded, and the thing, though being in its place, is not seen. Thus we have to say, proposition 3 is equivalent only to the following proposition:

Egocentric Language	Usual Language
3. As before.	3*b*. It is very probable that the thing exists during the interval Δt.

We find that proposition 3 is equivalent not to a proposition concerning the existence of a thing but to a sentence ascribing a probability to the existence of a thing.[10] We come to a similar result if we examine other examples. We find that normal propositions about the existence of things cannot be expressed in the egocentric language; this language can only express sentences about a probability for the existence of things.

This remarkable feature of the egocentric language is to be interpreted in the following way. The egocentric language confers existence only upon observed things, or, what amounts to the same thing, upon impressions.[11] Impressions are the basis of a probability inference directed to other things. A statement about impressions is therefore not equivalent to a sentence about physical things; it can only be equivalent to a statement conferring a probability upon a sentence about other things. The egocentric language dealing with impressions only cannot be equivalent to a language concerning the physical world.

[10] Strictly speaking, this is not an equivalence but a unilateral implication from the egocentric language to a probability statement about the realistic language (cf. our remark at the end of § 17).

[11] I do not mean by this that observed things and impressions are identical. But there is a one-one correspondence between them, and therefore the egocentric language can be formulated either for observed things or for impressions.

It can only be equivalent to a language conferring probability upon statements about the physical world.

Our investigation, therefore, confirms our thesis that the relation between impressions and the physical world is a projection and not a reduction. Impressions remain external elements of the world and cannot be considered as internal elements. The positivistic idea that this distinction is a matter of definition only, and that a projection may be changed into a reduction without any change of meaning, is not tenable. The egocentric language which would take the form of conceiving the physical world as a reducible complex of impressions cannot furnish propositions equivalent to propositions concerning the existence of physical things but only equivalent to sentences concerning a probability for the existence of physical things. The egocentric language is not equivalent to the physical language but only to a part of it; this is the part concerning the basis of probability inferences. It is precisely the part concerning physical things, given by the result of the probability inferences, which finds no equivalent in the egocentric language.

These results show that the positivistic conception of the problem of existence is no longer tenable. The positivistic conception that the question concerning the existence of the external world is a pseudo-problem is based on the idea that the physical language is equivalent to an egocentric language. For only in case of such an equivalence are we entitled to contest the uncertainty character of the logical process leading from impressions to external things; if this is nothing but an equivalence transformation of language, there remains no uncertainty as to the existence of external things. We see, however, that this is erroneous; there is no logical equivalence between statements about impressions and statements about external things—the latter are obtained by inductive inferences based on the former. There remains, therefore, always an uncertainty in this inference. As in the case of any statement concerning a special physical thing, so the general statement that there are physical things at all, that there is an external world, can be maintained with probability only. The degree of probability of the general statement is higher than that of any special statement; this is due to the fact that the general statement may be conceived as a disjunction of special statements, a case for which the rules of probability furnish a higher numerical value. But there is no reason to maintain that the general statement is certain.

That there is a remaining uncertainty may be made clear by the following consideration. We know that during a dream we have the feeling of the reality of the world observed, and we know that after awaking we are obliged to correct our conception—that we must acknowledge it was only a private world in which we lived. Can we exclude the case that a similar discovery will happen tomorrow with respect to the world of today? Can we ever ascertain, with no doubt remaining, that we are not asleep? Or are we sure that there will never be a third world, of a stronger reality still than the second, which stands to the second in the same relation as this to the first, the dream world? In denying such possibilities, we can never pass beyond a certain high degree of probability.

To these latter reflections it might be argued that our actual inferences to the external world start from a restricted class of impressions, limited by the impressions of today; and it might be argued that it is only this limitation which furnishes the uncertainty. If we could foresee all future impressions, we should know whether we should "awaken" some day; the statement of the existence of the external world would then be equivalent to the statement that there are no such impressions of "awakening" within the whole class. This argument is not valid, however, because even the knowledge of the whole class of impressions of a man's life does not furnish a basis from which we could infer with certainty the existence of external things. I do not admit that we can ever describe a class of impressions about which we may say that, if all impressions of my life are of this class, there is, with certainty, an external world. On the contrary: to any class of impressions described, even if it contains an infinity of impressions, we may imagine additional elements such that the enlarged class will lead to the conclusion that the world of the original class was a sort of dream world only. *All definable classes of impressions are of a type leading only to probability statements about an external world.* This is what we formulated as the nonequivalence of the realistic and the egocentric language; and this is what gives the reason for the uncertainty of our knowledge about the existence of an external world.

§ 17. Positivism and realism as a problem of language

With the reflections of the preceding section our inquiry about the difference of the positivistic and the realistic conception of the world has taken another turn; this difference has been formulated as the difference of two languages. This form of consideration, which has been applied particularly by Carnap, seems to be a means appropriate to the problem in question, and we shall make use of it for an illustration of our results.

The conception of the difference in question as a difference of language corresponds also to our idea that the question of meaning is a matter of decision and not of truth-character. If, in the preceding sections, we defended the idea that the positivistic conception of the world is not tenable, it was because we wanted to maintain that the positivistic interpretation of existence propositions does not correspond to our common language, or to that kind of meaning which we have to attach to our words if our actions are to be considered as justifiable in terms of our knowledge about external things. Our statement, therefore, belonged to the descriptive task of epistemology (§ 1), maintaining a deviation of the positivistic interpretation from the realistic language of knowledge as a given sociological phenomenon. If we proceed now to regard the differences of the positivistic and the realistic languages, we pass from the descriptive task to the critical task of epistemology; with this turn we consider meaning as a matter of free decision, and ask for the consequences to which each form of decision leads, and thus for the advantages and disadvantages which may be used to determine our choice if we ourselves want to make a decision.

In spite of our reference to a free decision, we should not like to say that the decision in question is arbitrary. Although such a characterization cannot be called false, it is a very misleading mode of expression. If we speak of the arbitrariness of language, we intend to express the fact that different languages may express the same ideas in spite of all differences of external form; and that, consequently, the choice of the language does not influence the content of speech. This conception has its origin in certain characteristics of common languages; it does not matter whether a scientist expresses his ideas in English, or French, or German, and thus the irrelevance of the choice of the language has become the very prototype of arbitrary decision. This conception presupposes, however, the equivalence of the languages in question. Only in the case of equivalent languages are their differences matters of convention. There are, however, other cases in which the languages are not equivalent; our consideration of the egocentric language led us to an example of this kind. In such a case the decision for or against one of the languages signifies what we called a volitional bifurcation. If we speak in such a case of an arbitrary decision, the word "arbitrary," therefore, is misleading; it suggests the idea that the decision in question is not relevant, does not influence our results. This, however, would be entirely erroneous.

If the languages in question are not equivalent, if the decision between them forms a case of a volitional bifurcation, this decision is of the greatest relevance: it will lead to consequences concerning the knowledge obtainable. The man who speaks the egocentric language cannot express certain ideas which the man with the realistic language may formulate; the decision for the egocentric language, therefore, entails the renunciation of certain ideas, and may, consequently, become highly relevant. We do not thereby say that the egocentric language is "false"; such a criticism would be a misunderstanding of the character of a volitional decision. It is rather the method of entailed decisions which we have to apply here; we can show that the decision for the egocentric language leads to a scientific system of a restricted character which does not correspond to the system constructed by the realistic language in its full extension.

Let us extend similar considerations to the general case of two languages. Using a symbolism corresponding to that of § 15, we will assume a domain I of elements as the basis of our language; let us assume, further, that statements i concerning these elements are absolutely or practically verifiable. With the latter term we include cases in which the statements i possess a high weight. There may be, in addition, a domain E of elements outside the domain I of elements; the elements of the domain E are in such relation to those of *I* that some verified statement *i* confers a determinate probability to a statement *e* about the domain *E*. This relation is not simply a one-one correspondence; to every *i*, there belongs a class of statements *e*, each of which is co-ordinated to *i* with a different probability (and conversely). Let us assume now two statements e_1 and e_2 with the following characters:

α) A determinate verified statement *i* confers on e_1 and e_2 different probabilities which are, however, not so high that they may be considered as practical truth

β) e_1 and e_2 differ with respect to predictions of facts happening outside the domain *I*

γ) e_1 and e_2 do not differ with respect to predictions of facts happening within the domain *I*

We will now introduce two languages; the narrower language may be defined by truth meaning in combination with the principle of retrogression, the wider language by probability meaning. The wider language will call the statements e_1 and e_2 different. The narrower language will also accept them as meaningful because they involve predictions for the domain *I*; this language, however, cannot acknowledge any difference between e_1 and e_2 because the predictions involved for *I* are the same, and all difference is based on a calculation of probabilities which are too low to serve as practical verification. Thus the narrower language calls the statements e_1 and e_2 equivalent. For this language, there is as much meaning in a statement as can be (absolutely or practically) verified within *I*; this language, therefore, may replace both the statements e_1 and e_2 by the statement *i*, if *i* is conceived as involving the same predictions for *I*, and call *i* equivalent to e_1 and e_2.

Of such a kind is the language of the positivist concerning the cubical world. In realistic language, e_1 and e_2 are two different hypotheses about the birds and *i* is the co-ordinated description of the pairs of shades. The restriction to the domain *I*, as basis, is due in this case to the physical conditions; a statement *e* is therefore excluded from absolute or practical verification. Whether a statement *e*, however, has a meaning different from *i* is not determined by the physical conditions but depends on the choice of the language. If we decide for physical truth meaning, we obtain the narrower language and are to call *i* and *e* equivalent; if we decide for probability meaning, we obtain the wider language and are to call *i* and *e* different.

We cannot forbid anyone to choose the definition of meaning he prefers. If he makes his decision, however, the previous considerations form a logical signpost for him. He may be right in saying that, as long as a hole in the walls is excluded, he cannot distinguish between the statements *i* and *e*; this is true if he decides for physical truth meaning, i.e., for the narrower language. What would be entirely false, however, would be an utterance from his side that *we* cannot differentiate between *i* and *e* as long as there is no hole in the walls. We can; this is because we may choose the wider language, based on probability meaning.

These considerations demonstrate a restricting quality of truth meaning. If the domain of basic elements is restricted, truth meaning leads to a restricted language, for which statements concerning elements outside the basis are meaningless unless they are conceived as equivalent to statements concerning elements of the basis. Probability meaning, on the contrary, is free from such restrictions; it may pass beyond the basis of the language.

Let us apply these results to the language of impressions. If the basic domain of the language is restricted to the impressions of one man, attainable by him during the stretch of his life, statements about things happening before or after his lifetime are meaningless except in so far as they are interpreted as being equivalent to statements concerning impressions of his lifetime. The two statements e_1 and e_2 used in § 16 concerning events after one's death are of this type; they possess the qualities α–γ and cannot be distinguished within this language. This is the decisive difficulty of positivism. The strictly positivistic language—thus we may call this language—contradicts normal language so obviously that it has scarcely been seriously maintained; moreover, its insufficiency is revealed as soon as we try to use it for the rational reconstruction of the thought-processes underlying actions concerning events after our death, such as expressed in the example of the life insurance policies (§ 16).[12] We have said that the choice of a language depends on our free decision but that we are bound to the decisions entailed by our choice: we find here that the decision for the strictly positivistic language would entail the renunciation of any reasonable justification of a great many human actions. The pragmatic idea that the definition of meaning is to be chosen in adaptation to the system of human actions, that it is to be determined by the postulate of utilizability, decides, therefore, against the strictly positivistic language.

To avoid these difficulties, positivists have attempted some generalizations of their language by an enlargement of the basis. Instead of the impressions of one man, they have considered the impressions of living beings in general as the basis. Such an expansion, however, contradicts the epistemological intentions of positivism which were to construct the world on the basis of one's own psychical experience; if this domain is once passed, there is no reason

[12] We may add that similar examples might be constructed for events situated before our lifetime, with the difference, however, that in this case the problem of action is not so directly concerned.

to stop just with the impressions of other people and to exclude other things. I should say that speaking of the independent existence of a table or of a stone seems much more permissible than speaking of the impressions of other people. Moreover, the expansion described does not suffice to solve all difficulties. There remain similar difficulties for events situated before the origin or after the expiration of mankind.[13]

Another expansion would be the introduction of a mixed basis. The basis determined by the impressions of our lifetime is defined by the postulate of physical possibility, i.e., by restricting meaning to impressions, the occurrence of which is physically possible. We might enlarge this basis by deviating from this postulate to a certain extent, admitting logically possible impressions situated at any time or any place in the world. In thus extending the domain of possible impressions throughout time and space positivists usually refuse to countenance expansions brought about by physical changes of the human body. It cannot be called logically impossible that the human body should become as small as an atom, or as large as the planetary system; the usual positivistic objections against the direct meaning of sentences about the elementary particles of matter refer therefore to physical possibility and not to logical possibility. But, if the case of logical possibility is once admitted for the spatiotemporal extension of the linguistic basis, it might be admitted as well for other extensions. It is true that we cannot forbid anybody to exclude the latter expansion and yet admit the former; we cannot see, however, much cogency in the construction of such a mixed basis. The arbitrary character of its limits becomes evident in some of its consequences: sentences

[13] This has been emphasized, with good reason, by C. I. Lewis, "Experience and Meaning," *Philosophical Review*, XLIII (1934), 125.

about events after our death are admitted as meaningful; sentences about the atom are prohibited, or reduced to sentences about macroscopic bodies. In spite of such scarcely justifiable qualities, this seems to be the basis which implicitly underlies most of the positivistic theories.[14]

It might be proposed to admit logical possibility to its full extent: to introduce a basis encompassing all kinds of logically possible impressions, including those which would occur with changes in the human body. This, we might suppose, would be the widest possible basis; with it we would presuppose nothing but the logical necessity of an impression basis—for that there is such a basis of impressions, that knowledge is conferred upon us through the medium of impressions, seems to be logically necessary. Or can we imagine that we may on some occasion get out of our private world?

This question, I think, is not to be answered in the negative—at least if the term "my own experience" is to have a meaning different from the purely logical term "basis of inferences." That there is such a private world is not a logical necessity but a matter of fact only, caused by the physiological organization of the human body. That I have to speak of *my* impressions, that I am separated from the impressions of other people, is by no means logically necessary. It is a matter of fact in the same sense as the people of the cubical world are bound to the interior of their cubical world. I could imagine other worlds in which impressions are not always bound together to the bundle "I"—worlds in which perhaps sometimes the ego splits into two egos which afterward unite again (cf. § 28). I can by no means maintain, with certainty, that all future expe-

[14] The refusal to admit physical changes of the human body finds its expression in Mach's struggle against atomism (cf. § 25).

rience will be of the same kind as present experience, will consist of colored figures and loud tones and resisting tactile sensations. This world may change in a way which we cannot imagine. Thus the statement, "Knowledge is bound to impressions as basic facts," is not absolutely certain.

It follows that the basis of all logically possible "impressions" is not the widest basis possible and would involve some restrictions; it seems, we must add, that a widest basis cannot be properly defined at all. To say, "All inferences about external things must start from elements of such and such kind" will never be permissible because we cannot define this "kind" in such a way that human beings are necessarily restricted to elements of the type described in order to have a basis of knowledge. Thus truth meaning will always lead to a restricted language, given any basis whatever.

The way to keep free from restrictions is pointed out by probability meaning: probability meaning, applied to any basis whatever, leads to an unrestricted language. This, it seems to me, is a decisive argument for preferring probability meaning. We may begin with a rather small domain of basic elements and construct upon it statements concerning elements of another domain without being obliged to borrow their meaning from statements about the basic domain. Thus probability meaning leads to the realistic language of actual science; we start from the rather small domain of our own observations and construct the whole world upon it. The positivistic postulate that the meaning of statements about this wider world is to be interpreted in terms of statements about the basic domain turns out to be not an obvious principle but the product of too narrow a conception of scientific language. This ambitious postulate is to be logically qualified as a proposal for a certain restricted language; there is, however, no reason for us to accept a proposal which involves the renunciation of a great deal of human knowledge. Our situation with regard to external things is not essentially different from that of the inhabitants of the cubical world with respect to the birds outside: imagine the surface surrounding that world to contract until it surrounds only our own body, until it finally, with some geometrical deformations, becomes identical with the surface of our body—we arrive, then, at the actual conditions for the construction of human knowledge, all our information about the world being bound to the traces which causal processes project from external things to the surface of our body. We may therefore apply the analysis of the cubical-world model to the case of the relation between impressions and external things. What was shown for the cubical world is that only physical truth meaning binds us to the domain *I* of given facts; if we accept physical probability meaning, we may pass beyond the domain *I* even if all observable facts are restricted to it. The same is valid for the relation of impressions to external things. Only if we confine ourselves to physical truth meaning are our sentences bound to impressions alone. If we accept physical probability meaning, we are not bound to this domain; our statements may pass beyond it and refer to external things. This is what the logical signpost states; we do not forbid anyone to decide for the definition of meaning he likes—but if he decides for truth meaning, such as do the positivists, we do not admit that he substantiates his decision by saying that a statement about external things, as distinct from statements about impressions, cannot be conceived as meaningful. The equivalence is valid only for his definition of meaning; there is another definition of meaning, however, based on the probability concept, which may differentiate between statements about external things and statements about impressions, even though it is not physically possible to extend the domain of observable facts beyond the domain of impressions.

A critical survey of the problem of impressions and external things therefore leads us to a confirmation of our refusal to accept the positivistic doctrine. The theory of the equivalence of statements about impressions and statements about external things originates from a too narrow conception of meaning; we are not restricted to this conception—and actual language has never been limited to such a narrow precept.

It may be proposed to formulate the relation of the positivistic to the realistic language in the following way. Since impressions furnish only probabilities for external events, a statement equivalent to a statement about impressions would be a statement concerning a probability of external events. If we introduce the name of statements of the second level for statements of the latter type, we may say that the impression language is equivalent to the second-level language of science. This would be a far-reaching change in the intent of positivism, since with this idea the existence of an independent realistic language not equivalent to the impression language is admitted. We might, indeed, agree with such a conception; we must add, however, that it can be carried through only in the sense of an approximation. There is, first, the difficulty that statements about impressions only imply probability statements about things but are not equivalent to such statements; the construction of the whole equivalent class of impressions would lead to difficulties similar to those described for the original positivistic conception. Second, the second-level language is, strictly speaking, not a two-valued language but once more a probability language, only of a higher level (cf. our criticism of impression statements in the following chapter and our remark on weights of higher levels in § 43). The interpretation indicated is, however, likely to be the best interpretation of positivism we can have: in the first approximation positivism is considered as equivalent to the language of science; in the second approximation positivism is considered as equivalent to the second-level language of science. The second approximation is much more exact than the first.

§ 18. The functional conception of meaning

If we now summarize the results of the present chapter, we find that it is the neglect of the probability character of the relations between impressions and external things which constitutes the fault of the postivistic construction of the world. The true-false conception of knowledge is valid only in the sense of an approximation; it must be applied, therefore, under careful control, and the consequences to which it leads must be interpreted in full consciousness of the merely approximative character of the presuppositions. Positivism, therefore, if it is to be considered as a permissible conception of the world, must be conceived as an approximation; only in this sense may it become of scientific value.

In this sense it is indeed frequently used and with success. If a new scientific theory is started, we imagine a set of impressions which if observed would make the theory highly probable; we say, then, that we understand the theory. If its truth is in question, we imagine another set of impressions which if observed would make the theory highly improbable; we say, then, that we understand how a refutation of the theory would run. The positivistic method thus provides us with a good intuitive representation of the theory; but it does no more.

In this process of making a theory intuitively clear it may also be permissible to supersede the postulate of physical possibility and to introduce imagined impressions which are logically possible only. If this expansion is not always consistently carried through, if some logical possibilities are admitted and others rejected, we shall not oppose such a mixed basis; it may even be advisable to refrain from drawing too narrow limits. We read Gulliver's voyage to the Lilliputians and picture with pleasure impressions we should have in this miniature country, although it is not physically possible to go there. Reading Einstein's theories, we imagine a man who sets his watch right by the arrival of light rays with a super-astronomic precision; although this is not physically possible, it may be a good representation of Einstein's definition of simultaneity. We fancy rotating atoms and jumping electrons as though we could see them with a microscope, and that may be a good help for understanding Bohr's theories. The physicists have shown that we must be very careful in such constructions, that some of the tacitly assumed conditions of our macroscopic world are no longer valid for sub-microscopic dimensions; but in picturing a world which is constructed half by the postulates of physical laws, half by suppositions extending beyond physics, we may understand some essential features of the world which had previously escaped our notice and advance toward an intuitive understanding of theories which would otherwise remain in the mists of abstraction.

We must not forget, however, that the set of impressions fancied is not equivalent to the intension of the theory in question. Assuming this is just the illegitimate consequence to which the neglect of the probability character of knowledge leads. It means disregarding the fact that every describable set of impressions, if observed, furnishes probability only for physical statements. It means overstraining the bearing of approximative concepts and deducing from them consequences for which the limits of the approximation do not hold. It means restricting one's self to an intuitive representation—the occurrence of some determinate impressions—instead of exhausting the meaning of the whole sentence. It is not, as positivists pretend, the only admissible conception of meaning but an oversimplified theory of meaning.

The origin of this theory of meaning, it seems to me, is

to be found in the idea that the meaning of a sentence is something which may be pointed out, which may be seen and known. This "something" is constructed by positivism in the set of impressions belonging to the sentence. What we obtain in this way, however, are only images, associated representations. It is a psychological conception of meaning which positivism maintains—based, however, on some metaphysical remainders taken over from traditional philosophy—from a substantial conception of meaning. It is this deep-rooted misconception from which the positivistic theory of meaning originates.

The meaning of a proposition is not "something"—there is no question at all of the form, "*What* is the meaning?" A proposition has meaning—that is, a proposition has certain qualities; but it does not have a co-ordinated something which *is* the meaning. We had better say: a proposition *is significant*—the substantival term, "has meaning," is always to be understood in the sense of the adjectival term "is significant." This corresponds to our usage of words in the two principles of the theory of meaning which define not the use of the term "meaning" but that of the term "has meaning." The first denotes under what conditions a proposition has meaning, the second denotes under what conditions two propositions have the same meaning; this is all we need—we need not know what the meaning is.

To understand a proposition is the desire of every good-intentioned scholar, and it appears perhaps a heartless radicalism if we maintain that there is no understanding in the sense of "knowing the intension." What we call understanding, however, is nothing but producing associated images, representing some effects connected with the sentence, forming an intuitive representation. We do not intend to forbid this, certainly. We are convinced that this is a very good and fertile way of working in science, that intuitive images may make thinking distinct and creative, that it is perhaps just these associations to which is due the intense joy combined with all productive and reproductive scientific thinking. What we object to, however, is the identification of the associated images with the meaning of the propositions, and the substitution of an intuitive representation for the full and complete intension. In other words, we refuse to deduce the meaning of meaning from psychological processes.

Thinking works in a tunnel; we do not see intensions, contents. Propositions are tools with which we operate; all we can demand is to be able to manipulate these tools. The darkness of the tunnel may be lighted by the searchlights of intuitive images fitfully appearing and wandering. Let us not confound blurred images with the full class of operations for which the tools are good.

Reference to impressions is permissible in the sense of an intuitive representation—if we accept this, however, we may accept other representations as well. The realistic conception of the world possesses images of this kind as well as the positivistic conception; and I do not see any reason why these conceptions should not be permissible in the same sense as positivistic images. Positivists have attacked realism in pretending that it is meaningless to imagine external things which we do not observe, and then have insisted that the only permissible interpretation of propositions about external things is to realize the impressions we should have when the things were observed. This, it seems to me, is the attack of one metaphysician against another; it cannot be the task of scientific philosophy to decide for one side in this struggle. An unprejudiced analysis of scientific propositions shows that the positions of positivism and realism are both rooted in the psychological sphere and that the concept of meaning should be freed from all such psychological components if it is to correspond to the practice of thinking.

Meaning is a function of propositions; it is that function which is expressed in their usefulness as instruments for our actions upon the world. Meaning is not a substantial something attached to a proposition, like "ideas" or "impressions," but a quality; the physical things called "symbols" have a certain function as to operations on all other things—this function is called meaning. It is this functional conception of meaning only which opens the field for the introduction of the concept of probability into the theory of meaning. Probability meaning, as we defined it, must be considered within the framework of this functional theory. It seems to me that only this combination with the probability theory can provide the functional theory of meaning with the tools necessary for a satisfactory theory of scientific propositions, a theory adapted to the actual procedure of science. This is what is shown by the analysis of the relations between impressions and the external world.

CHAPTER III

AN INQUIRY CONCERNING IMPRESSIONS

CHAPTER III

AN INQUIRY CONCERNING IMPRESSIONS

§ 19. Do we observe impressions?

The foregoing chapter was based on the presupposition that impressions are observable facts. We introduced them because we found that physical observations, even of the most concrete type, can never be maintained with certainty; so we tried to reduce them to more elementary facts and arrived at impressions as the immediately given facts. It may be doubtful, we said, that there is a table before me; but I cannot doubt that at least I have the impression of a table. Thus impressions came to be the very archetype of observable facts.

This train of thought is of convincing power, and there are not many philosophers who have been able to resist it.[1] As for myself, I believed in it for a long time, until I discovered at last some of its weak points. Although there is something correct in these reflections, it seems to me now that there is something in them which is essentially false.

[1] If I am to give some names among this exceptional group, I have to mention first Richard Avenarius, whose struggle against the "introjection" of the psychical phenomena and for a "Restitution des natürlichen Weltbegriffs" is the first clear refutation of a standpoint which materialists at all times had already attacked with much ardor but with insufficient means (Avenarius, *Der menschliche Weltbegriff* [Leipzig, 1891]). Recently, Watson in his behaviorism (*Behavior* [New York, 1914]), and Carnap and Neurath in the behavioristic turn they gave to the Vienna positivism (*Erkenntnis*, III [1932], 107, 204, 215) developed similar ideas and in a much more easily accessible and therefore more convincing form. My following exposition, though related to behaviorism, differs however in some respects from it (cf. § 26). Pragmatists also have resisted the positivistic dogma; Dewey, in *Experience and Nature* (Chicago, 1925), gives a very clear refutation of the idea that impressions or sensations are observable facts. Cf. also the very convincing form of behaviorism developed by E. C. Tolman, "Psychology versus Immediate Experience," *Philosophy of Science*, II, No. 3 (1935), 356.

I cannot admit that impressions have the character of observable facts. What I observe are things, not impressions. I see tables, and houses, and thermometers, and trees, and men, and the sun, and many other things in the sphere of crude physical objects; but I have never seen my impression of these things. I hear tones, and melodies, and speeches; but I do not hear my hearing them. I feel heat, and cold, and solidity; but I do not feel my feeling them. It may perhaps be answered: It is true that you do not see your seeing, or hear your hearing; but you sense it in another way, with an "internal" sense which furnishes a direct sensation of impressions corresponding to sensations of external objects furnished by the other senses. But, though this conception of an internal sense has been maintained since Locke by many philosophers, I confess I do not find such a sense within myself.

I do not say that I doubt the existence of my impressions. I believe that there are impressions; but I have never *sensed* them. When I consider this question in an unprejudiced manner, I find that I *infer* the existence of my impressions. To show the structure of this inference, let me give an example taken from physics.

Electricity is an entity which has never been observed by any man. We cannot see it; we infer it. We see copper wires and observe that these wires have different qualities without a visible change in them: sometimes, if we touch them, we feel a shock, and sometimes not; sometimes a lamp connected with the wires lights, sometimes not. To justify this difference of observable facts connected with copper wires we assume that there is an unobservable thing in them which we call electricity.

Of the same type, it seems to me, is the inference leading to impressions. We experience that the things we observe have different qualities, just as have the copper wires. The main difference is given by the two worlds of dreams and wakefulness: sometimes the things we observe remain for a long time, sometimes only for a short time; sometimes they show constant and persisting qualities, sometimes they offer curious and surprising aspects and combinations. To explain this difference, I introduce the distinction between the physical thing and my impression of the thing; I say that usually there are both physical things and impressions within me but that sometimes there are impressions only without corresponding physical things. The responsibility for the confused and curious things is thus taken away from the "external" things and transferred to another thing called "I." But with this conception the world is doubled; we maintain that also in the regular case of well-ordered things there is the duplicity of external things and my impressions. We need this assumption to justify the explanation that in the case of the confused world one of the two worlds, the external world, is dropped. The distinction between the world of things and the world of impressions or representations is therefore the result of epistemological reflection. It is well known how long a time it took for mankind, in its historical development, to discover this distinction; even today primitive peoples show a confusion of both worlds—they take dreams for realities and substantiate actions of the waking world by experiences they had in dreams (cf. § 25). There is no direct awareness of impressions or representations; we must learn to infer whether the things we observe are "real" or if they are only "apparent," this term meaning that there are processes in my body alone which are not accompanied in the usual fashion by physical things.

I do not say that this reduplication is a false theory; on the contrary, it is a very good one. It explains many facts such as the difference between the image of the concave mirror and the material table, or between the flash of light produced by a stroke with the fist and the flash of light produced by a lighthouse. In these cases there are external things of quite different character, though *I see* the same external things; and the duplicity theory explains this by assuming that different external objects may produce the same internal process within me. Thus again the distinction between the external thing and the internal process of sensation furnishes a reasonable explanation. This theory, therefore, is as good as any physical theory of a similar kind; but it is a theory and not an observation.

This abstract character of impressions has perhaps been obscured by a prevalent attention to the sense of touch. As for optical impressions it is obvious that I do not see them; but for the tactile impressions it may appear permissible to say that I feel them. This, however, seems to me to be a confusion due to a certain peculiarity of the sense of touch. If we touch an object, we localize it at a spatial point which is situated on the boundaries of our body and not at a distance from the body, as in seeing. We can therefore say that the object we feel is in our body, and thus the idea arises that we feel an impression. But in touching we always feel things. If we slide our hands along the edge of a table, we feel the table in the same sense as we see it with our eyes; blind men, who have had more practice than we, know this and are accustomed to attach the conception of sensing external things to their experiences of touching.

The matter is still more complicated by the fact that in certain cases the object which we sense may be a process occurring within our body. This is the case when we feel pains or hunger. But what we then feel is an occurrence in the same sense as when we see an object with our eyes; just as we see our body, we may feel it. That feelings like these have the character of a sensation is shown by the fact that they always appear accompanied by a definite localization within our body. We feel headaches in the head, hunger in the abdomen, an overstrained muscle in the leg; and there is a location also if we sense some feelings spread all over the body, like the feeling of fatigue. We are entitled to say that in such cases we sense the inner state of our body; but then there is this object alone, just as in the case of an optical sensation of a distant thing. A sensation of a sensation never occurs; there is only one sensation, its object is an external thing, or a state of our body, and that there is a sensation is not observed but inferred.

What is given are things, or states of things, including states of my body—not impressions. The cause of this confusion of an inference with an observation is to be found, to a certain degree, in the fact that given things have certain qualities which, as investigation shows, are not due to them, or not to them alone. Things are blue, or red, or warm, or hard; but science demonstrates that these qualities do not belong to the external things. To state this more precisely: Science shows that things have these qualities only when they enter into a relation with our body and not when they simply act on one another. When a blue body is put before the objective of a camera, it acts upon the film in the camera; but, if we try to understand this relation, we have to ascribe to the "blue" body the quality of emitting electrical oscillations which have no similarity to the color "blue." When a hot body is put into cold water, the water starts bubbling and fizzling and betrays in this way the occurrence of some mechanical energy which has nothing in common with the quality "hot." It is thus demonstrated that certain qualities are not qualities of the external thing alone but of the interaction between the external thing and our body. Such qualities are rightly called secondary qualities. Interactional qualities of such a specific kind may appear also through the mutual combination of external things without interference of the human body. In general, light rays do not change the bodies they strike; but, when they fall upon a photographic plate, they blacken it and may draw the silhouette of an intermediate body on the plate. Thus light rays possess a "power to draw" not as a quality of themselves in isolation but as an interactional quality occurring only in combination with certain other things. If this other thing is the human body, the interactional quality acquires a special importance; it is this kind of interactional quality which is called secondary quality, according to the traditional philosophic usage.

It is to be kept in mind, however, that the secondary qualities are *qualities* of things, not *things*. The confusion of this difference, the illegitimate objectivization of qualities, is one of the reasons for the false conception that impressions are observed. Philosophers talk of "the blue" which they observe, of "the hot," of "the bitter"; but this is an abuse of words. We never see "the blue," but blue things; we never taste "the bitter," but bitter things. Things as they are given appear provided with certain qualities; so we had better avoid expressions like "We observe these qualities," and replace them by "We observe things having these qualities." The false expression that "we observe qualities," together with the right idea that these qualities are due to a co-operation of our body, leads to the conception that we observe impressions. This seems to be the psychological origin of the untenable observation theory of impressions. Critical analysis replaces it by an inference theory of impressions.

The abstract character of impressions is indicated also by the way in which we describe impressions linguistically. There are no words denoting impressions. There are words for the secondary qualities; but no words exist for impressions as events in the whole. We describe an impression by denoting a thing which may produce such an impression. We say: "I had the impression of a red square," or "I had the impression of a flash of light." What are the things denoted here? A red square is a piece of red paper or of other material square in form; and a flash of light is a quantity of light as produced by lightning or by lighthouses. We add to such words the term "impression of . . . . ," and characterize in this way the impression. But this is an indirect way of description; we are obliged to employ it because the corresponding words of daily language concern only observable things and not impressions.

§ 20. The weight of impression propositions

The result of the foregoing section may be stated in the form that impression propositions are indirect, not direct. It is a great mistake to believe that proceeding from observation statements of physics to impression statements is a movement from "not wholly direct" statements toward "direct" statements, or at least toward "more direct" statements. The converse is true; this way leads to "less direct" statements, impression statements being the result of an inference and not of observation. The maximum of "direct character" is with the observation statements; from these there is one way of inference leading to the indirect propositions of physics, and another way of inference leading to the indirect propositions concerning "my impressions."

But, if we now proceed to analyze the weight belonging to indirect propositions of these two kinds, we find a remarkable difference. The weight of the indirect propositions of physics is inferior to the weight of observation propositions; this is due to the fact that the indirect propositions of physics have a surplus meaning as compared with the observation propositions. The indirect propositions concerning impressions, however, have less meaning than observation propositions, and therefore they have a superior weight. As this inverse behavior in respect of weight is a feature of very high importance, it must be explained in detail.

We have pointed out that an impression proposition is formulated in language by reference to physical objects which produce this impression. This is an essential feature because there are no other means to describe an impression. The description, however, is not performed by pointing to one object alone; we add some other objects which would produce the same impression. If we say, "I had the impression of a flash of light," this reads: "I had an impression such as is produced by the beam of a lighthouse, *or* by a flash of lightning, *or* by a blow with the fist on my eye." Impressions are therefore characterized by a disjunction of physical objects. The occurrence of this disjunction produces the diminution of intension; we shall point out later how this is performed. Now we must consider more precisely the disjunction occurring here.

It is not always necessary to enumerate all the terms of this disjunction. This can be avoided by the use of the concept of similarity; and we must show how this is to be done.

The objects we sense are not always different; some of them are very similar. If I look at this table, and then look five minutes later, the second table is similar to the first one; I usually even say that it is the same table. This is a bit imprudent, as far as it concerns only what I see just now; I had better say, "The table Number 1 is in the relation of similarity to the table Number 2." Whether this similarity is to be interpreted as identity of the physical objects depends on a number of other circumstances. The table Number 3, which I saw in another room, is also in the relation of similarity to the two other tables, but it is not physically identical. This is of course not directly observed, as little as the physical identity in the other case is observed; it is inferred from some other relations between other things. Thus the relation of physical identity expresses a complex of elementary relations. The primary relation is that of similarity; and our observation statements consist primarily in maintaining that the relation of similarity is valid between several things.

By this means we can characterize the terms of the disjunction demarcating an impression. For example, we can speak of "an impression produced by the beam of a lighthouse, or by another physical object which stands in the relation of similarity to such a searchlight." It is to be noted that this concept of similarity is different from "physical similarity," since a light ray would be similar, in our sense, to a blow with the fist on the eye. Our similarity is what philosophers call "similarity of impression"; but it is to be remarked that we need not introduce the term "impression" to characterize this similarity—we can define the relation by pointing out a quality of things as we see them. We could say that it is a quality of things as a primitive man sees them, i.e., a man who was never perverted by philosophical analysis. Let us call this relation *immediate similarity*.[2]

Since we did not employ the term "impression" in the construction of our disjunction, we can drop it and express our statement in the form, "There is a thing a_1 or another thing standing in the relation of immediate similarity to

[2] The importance of the similarity relation for the logical construction of basic statements was first pointed out by Carnap in his *Der logische Aufbau der Welt* (Leipzig and Berlin, 1928).

a_1." Let us call this statement the *similarity disjunction*. From its form as a disjunction, it becomes obvious that a diminution of intension is performed; to state, "There is the thing a_1 *or* another thing," states less than, "There is the thing a_1." Our disjunction, however, is not yet sufficiently extended; we must expand it by a further term, and in this expansion the word "impression" will enter. The term to be added concerns the phenomenon of the dream.

If we "see the thing a_1 in a dream," there is no physical thing at all, but only an impression as it would have been produced by the thing a_1, or another thing similar to it. It is true, we do not know this while dreaming; but we know it afterward, and therefore we must take account of this case by adding this possibility to our disjunction.

The impression is my own internal state as it is produced by a_1, or a thing similar to a_1. To understand this completely, we ought to give an explanation of the term "my own"; however, this may be postponed to a later section § 28). Independently of this explanation, we may say that the term "impression" is defined by means of the concept of immediate similarity. But, although it is defined with reference to the object a_1 or to similar objects, the statement that there is, besides the object, an impression as an internal state of my mind, adds something to the statement about the object alone. Thus stating, "There is the object a_1, or an object similar to it, *and* in addition a corresponding impression," would be an increase of intension.

We add, however, the new term not in the form of a conjunction but in the form of a disjunction; so we obtain a further diminution of intension, compared with the similarity disjunction so far considered. The new statement reads: "There is the thing a_1, or a thing similar to a_1, or there is no observed physical thing, but only an impression as it would have been produced by the thing a_1." We call this statement the *longer similarity disjunction;* the previously constructed disjunction may be called *shorter similarity disjunction*, if it is distinguished from the longer one.

Let us denote by $S'(a_1)$ a thing similar to a_1; the sign $S'(a_1)$ thus already means a disjunction, constructed of all things similar to a_1. By $I'(a_1)$ we denote an impression of the type produced by a_1. The sign $\vee$ reads "or." Then our two disjunctions have the form

Shorter similarity disjunction: $a_1 \vee S'(a_1)$
Longer similarity disjunction: $a_1 \vee S'(a_1) \vee I'(a_1)$

We shall call statements of such a kind *basic statements.* After having construed their logical form, it is easy for us to show that they lead to a higher weight. This is due to the diminution of intension; the calculus of probability expresses this relation by an inequality[3] stating that the probability of a disjunction is greater than (in exceptional cases equal to, but never smaller than) the probability of each of the single terms of the disjunction. This is why the transition to basic sentences involves an increase of the weight; we need no "intuition" to prove this, or any "immediate knowledge of the certainty of the given" —we need nothing but the rules of probability. The longer similarity disjunction has a still higher weight than the shorter one.

We may construct a third form of basic statement by adding the assumption that there is also an impression in the case of the first terms of the disjunction. That is, we also state the occurrence of the impression in the case of the existence of the physical object. This combination may be called the *impression form;* it reads in symbols

Impression form: $a_1 . I'(a_1) \vee S'(a_1) . I'(a_1) \vee I'(a_1)$

[3] Cf. the author's *Wahrscheinlichkeitslehre* (Leiden, 1935), p. 97, eq. (13).

The introduction of the impression into the first terms means a diminution of the weight; but we may conceive it as highly probable that there is always within myself an internal process when I see a thing, and so the weight of the impression disjunction is not much less than the weight of the longer similarity disjunction. According to a rule of logistics,[4] the disjunction occurring in the impression form is equivalent to the term $I'(a_1)$; so we get the simple expression

$$\text{Impression form: } I'(a_1)$$

We see that the third form of basic statements is nothing but the statement that there is an impression of the type produced by the thing a_1. This is the form usually employed by positivism.

We add some examples. A shorter similarity disjunction is expressed in the statement: "There is a searchlight or a thing similar to it." Things of the latter kind would be a flash of lightning, or a blow of the fist. The transition to the longer similarity disjunction would be performed by adding "or I have only the impression of a searchlight." This would include the case that I am perhaps dreaming while stating the sentence. If it appears unjustified to anyone to call a blow of the fist a thing similar to a searchlight, he may cross this blow out of the shorter similarity disjunction and include it in the term $I'(a_1)$ of the longer one. This is a matter of definition only. The transition to the impression disjunction would read: "There is an impression of the type as produced by a searchlight." The latter statement, though of a rather high weight, is not quite as certain as the statement using the longer similarity disjunction; but the difference of degree is very small.

The weight obtained for the longer similarity disjunction

[4] Cf. *ibid.*, p. 27, 4*c**.

must now be given closer consideration. Is it equal to absolute certainty? Positivists and other philosophers have maintained this idea; for them impressions are indubitable facts, and they emphasize that just on this account impressions form the very basis of our knowledge of the external world. Our refusal to accept impressions as observable facts must influence this conception; we have to enter into an independent investigation of the weight occurring here. The guiding principle in this inquiry will be our interpretation of impression sentences as "similarity disjunctions."

We may take for granted that sentences of the kind, "There is a flash of lightning," are not absolutely certain. The increase of the weight toward certainty, if it comes about at all, must be performed by the introduction of the "or." Let us ask, first, whether the rules of probability can teach us something about this question.

There is a principle of probability stating that a complete disjunction $A \vee \bar{A}$ (i.e., A or non-A) has the degree of probability 1. Incomplete disjunctions have, in general, a smaller degree of probability; it is not excluded, however, that they have the probability 1. Now it is obvious that the similarity disjunction is incomplete. This must be the case because otherwise it would state nothing; to say, "There is a flash of lightning, or there is not," would be an empty assertion and could not furnish a basis suitable for information about facts. It follows that the rules of probability do not teach us anything about the question of the certainty of the similarity disjunction; they leave the question entirely open.

We must, therefore, look to other reflections as a guide to an answer to our question. We can obtain an answer if we consider the possibility of a later refutation of a basic sentence. For this purpose we must notice the meaning of the relevant terms.

If we say, "There is a flash of lightning, or an immediately similar object, or nothing but an impression of this type," the description is furnished by means of the physical thing "flash of lightning." This is because this term of the disjunction defines the other ones; the immediately similar objects are determinate only because they are referred to the flash of lightning. So too is the impression. Now the term "flash of lightning" denotes an object which has been formerly seen; the basic statement, therefore, gives a comparison between a present object and a formerly seen object. We admit that this comparison does not presuppose that the formerly seen object really was a flash of lightning, in the physical sense of this word; it is sufficient that it was an object which *I called* a flash of lightning. But this restriction does not influence our result that the comparison concerns both a present and a formerly seen object. Such a comparison, however, makes use of the reliability of memory and so is not absolutely sure. It turns out, therefore, that a basic statement is not absolutely certain.

The objection may be raised that a comparison with formerly seen physical objects should be avoided, and that a basic statement is to concern the present fact only, as it is. But such a reduction would make the basic statement empty. Its content is just that there is a similarity between the present object and one formerly seen; it is by means of this relation that the present object is described. Otherwise the basic statement would consist in attaching an individual symbol, say a number, to the present object; but the introduction of such a symbol would help us in no way, since we could not make use of it to construct a comparison with other things. Only in attaching the same symbol to different objects, do we arrive at the possibility of constructing relations between the objects; but in the distribution of the symbols the elementary comparison is then already performed. It is the function of the basic statements to formulate these elementary comparisons, under the viewpoint of immediate similarity; this is why basic statements can be used as a basis for further inferences.

We see that the conception of basic statements as absolutely certain propositions is untenable. This conception disregards the fact that basic statements never concern the present object only, but formerly experienced objects also —a feature which is essential to basic statements.

Our analysis of the weight of impression statements leads us to a psychological explanation of the theory which determines the positivist to believe in the character of impressions as elementary observational facts. The passage to less doubtful propositions is erroneously taken as the passage to more intuitive propositions. This conception is suggested by an analogous process for concepts of a higher level. Passing from "There is an electrical discharge from a cloud to the ground" to "There is a flash of lightning" is a transition to a more certain proposition and, jointly, to a more intuitive one. Passing from "There is a flash of lightning" to "I have the impression of a flash of lightning" is a transition, once more, toward a more certain proposition, but to a less intuitive one. Whereas the line of certainty permanently ascends in this transition, the line of intuitiveness ascends first, and later on descends, with a maximum on a certain middle level. We may be allowed to symbolize this idea by the diagram of Figure 3, although we do not intend to make proposals as to a practicable measurement of the degree of intuitiveness. It is the confusion of both lines which causes the positivistic conception of the immediate character of impressions—a theory which breaks down before the criticism of an unprejudiced psychological examination.

The higher degree of certainty co-ordinated to the impression statement is due to its character of being a disjunction. A disjunction does not lead, however, to an intuitive "more general thing"; the generalization is expressible in the terms of language only, but is not accompanied by a corresponding intuitive process. If "impression" is not identified with the inner process inferred but not observed, we should be obliged to interpret it as such a "thing defined by a disjunction," a thing, for example, which is either the flash of lightning or a thing similar to it. We cannot imagine such a "general thing"; what we see are always particular things, including qualities which perhaps are not objectively justified. We see the image in the mirror as a bodily thing; if we know this observation to be

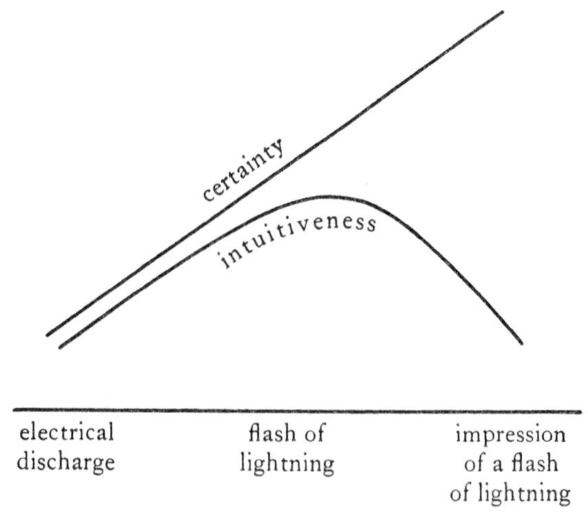

FIG. 3.—Transition from higher physical statements through observation propositions to impression propositions.

doubtful, we may reduce the intension of our statement by adding an "or," i.e., by saying, "There is either a bodily thing or only a bundle of light rays similar to it"—but we cannot see a "more general thing" such as would correspond to this disjunction. Positivism, with its conception of impressions as intuitive objects, has fallen a victim to the old metaphysical tendency to replace linguistic processes by intuitive entities. We cannot admit, however, that the nominalistic dissolution of conceptual realism, elsewhere the genuine tendency of the positivistic program, has to be stopped in face of the problem of the basic elements of knowledge.

§ 21. Further reduction of basic statements

Our conclusion concerning the uncertainty of basic statements raises the question whether we may carry on the method of reduction and arrive at statements of another kind which will be absolutely certain. This is to say that the direction toward certainty may be pursued a stretch farther; it will be no objection to this procedure if we admit that we perform this by reflection and not by an analysis of what is "immediately given."

Reflections of this kind may be substantiated by the fact that a direct comparison between a previously seen object and a present one is not possible. It is true that a basic statement gives a comparison between two objects and not a simple noting of one object; but what are compared are not objects at different temporal positions. When we see the present object, we no longer see the previously seen one; so we cannot compare them. Instead of the previously seen object we have only an image of it, furnished by memory; thus what we actually compare is a recollection image, on the one hand, and an object, on the other.

What is a recollection image? We know, in having such

an image, that, though we have the feeling of seeing an object, there is no object at all but only an internal process in the mind which we call an impression. But this is known only, not seen. What we see is not the impression but an object; and thus there is no other means of describing the impression than by describing the object which we have the feeling of having seen. This object is called a recollection image. The word "image" is to express that we do not believe in the reality of this object but that it is a representative of the original physical object. It would not be right, however, to say that the recollection image is "in" my head. In my head there is an internal process which I do not directly observe. The image seen is outside my head, in the place of a physical object although we know that there is no object at all.

Returning to our reflections concerning basic statements, we must admit that the comparison there spoken of is not performed directly but only by means of the intercalated recollection image. The comparison is divided into two processes: a comparison between the present thing and the recollection image and, second, a comparison between the recollection image and the previously seen thing. Now only the first comparison can be directly performed; the second has the character of a hypothesis: it is the assumption that the present recollection image is similar to the previously seen object. This is what is called the assumption of the *reliability of memory*.

We see that the analysis of this psychological process may indeed be interpreted as justifying the contention that our basic statements are not final elements but are capable of a further reduction which leads to new basic statements. Only the comparison between the present thing and the recollection image is a basic statement, properly speaking; the comparison between the recollection image and the previously seen object is performed by an inference, not by observation. Let us call the first comparison a *basic statement in the narrower sense*, whereas our former basic statements, combining both comparisons, may be denoted as *basic statements in the wider sense.* We are to say, then, that basic statements in the narrower sense contain only comparisons between present things. Basic statements in the wider sense are indirect sentences, based on basic statements in the narrower sense.

Before turning to the question of the certainty of the new basic statements, we must investigate the transition from basic statements in the narrower sense to basic statements in the wider sense. We said that this transition depends on the presupposition of the reliability of memory. This demands a more precise formulation.

Imagine that there is a certain confusion in our memory so that recollection images seen today, although caused by green bodies seen yesterday, are similar to red bodies seen yesterday, whereas the recollection images seen today, although caused by red bodies seen yesterday, are similar to green bodies seen yesterday. It would be impossible ever to discover this confusion because the comparison cannot be performed; we cannot directly compare a thing recollected today with a thing seen yesterday. So the supposed confusion would be meaningless, according to our definition of meaning. If the hypothesis of the reliability of memory should state that this confusion does not happen, the hypothesis would be a pseudo-statement and not worthy of further discussion. But we need not interpret the hypothesis in such a naïve manner; we can give it another interpretation which leads to a verifiable content (cf. also § 27).

To show this, let us introduce a method by which the recollection images are eliminated. It is true that, when we call a certain object a table, we compare it to a recollection image called forth by the word "table"; but we might employ another method. For this purpose, we might make use of our collection of specimens which contains specimens of all things, together with their designations (§ 5). When I say, "This is a table," I would then compare the object denoted by the word "table" in our collection of specimens to the object in question; so the recollection image would be replaced by a specimen taken from our collection, and the comparison would involve two physical objects but no recollection image.

We can say now what reliability of memory is: Memory is reliable when the method of recollection images leads to the same basic statements, in the narrower sense, as the method employing the collection of specimens. With this procedure the reliability of memory is defined in a testable way; it is the way which we actually use whenever the reliability of our memory is in question. If we doubt that our recollection image of a certain thing is right, we procure a new impression by looking at the thing. Sometimes the control is made by means of scientific textbooks and dictionaries; as these books do not furnish direct impressions but only definitions of the words, this procedure is to be conceived as the reduction of a recollection image to other recollection images of higher reliability.

In actual thinking the described strict method of comparison by means of the collection of specimens cannot be carried through on account of its technical complication. It is replaced by the function of memory. The reliability of memory can be controlled, as we have seen; this control, however, can only be performed in some special cases. For the other cases, we make use of an induction, supposing that memory is reliable also when it is not controlled. This hypothesis, however, lowers the certainty of the results.

Basic statements in the wider sense, therefore, are less reliable than basic statements in the narrower sense. The first result of our inquiry, therefore, is a confirmation of the idea that our former basic statements are not absolutely certain.

The transition from basic statements in the narrower sense to basic statements in the wider sense obtains a very simple form by means of the described hypothesis of the reliability of memory. If any basic statement in the narrower sense is given, we have only to replace the term "recollection image of the previously seen object" by the term "the previously seen object" and thus obtain the corresponding basic statement in the wider sense. The transition is performed, therefore, if we drop the reference to the recollection image and give to the basic statement the common form of a comparison between objects at different positions in time.

The transition in question contains a further hypothesis which we must now point out. It is the presupposition that objects which stood in the relation of immediate similarity, at a former observation, stand in the same relation when they are observed later on. We will call this idea the *hypothesis of the constancy of the perceptual function.* We must show how this assumption can be examined.

This examination can be performed by means of our collection of specimens. In this there are several objects bearing the names "flash of lightning," "beam of a lighthouse," "blow of the fist on the eye," etc., which show, in simultaneous comparison, the relation of immediate similarity. If we regard the same objects on the following day, we find that they still show the same relation. This is what is meant by constancy of the perceptual function.

Now it is obvious that this constancy will not be shown by all objects. It depends, in the common phraseology, on the physical constancy of the objects; if the objects change, the perceptual relations change. A tree seen in summer may be immediately similar to the color named "green" in our color table, whereas in winter it is immediately similar to the color named "white"—because snow fell in the botany department of our collection of specimens. But there are objects which do not change; more precisely formulated: If the objects are under certain observable conditions, they do not change. It is a matter of experience to find out these conditions. But if we have found them, we believe in the constancy of the similarity relation. This is not only a presupposition concerning the existence of invariant physical objects. It might happen that two objects show constantly no difference in respect to physical reactions of all possible kinds but that they look similar on one day and different on another day. The physical reactions of which we speak consist in chains of happenings, the results of which are observed by us; for this observation we may presuppose the constancy of the perceptual function and arrive at the result that the original objects did not change physically. But the direct observation of the objects may show that the objects are not similar, although they were before. Two rectangular sheets of white paper may look similar on one day, whereas the next day they do not look similar, and instead one of them may look similar to a circular sheet of paper—although an examination by means of rules and meter bars shows that the paper still has the rectangular form. The similarity relation depends not only on the physical qualities of the objects but also on a certain constancy of the sensational processes in the human body; this is what we call the *constancy of the perceptual function.*

This constancy is presupposed also in the transition from basic statements in the narrower sense to basic statements in the wider sense. It is contained in the use of certain words denoting, in current language, impressions. We say "the impression of a white rectangle" and suppose in using this term that all objects which furnish this impression, i.e., which are immediately similar to a rectangular sheet of white paper, will also be immediately similar later on. Without this presupposition, the use of words as we employ them would be ambiguous; we should always have to add a time index, such as, "the impression of a flashlight as it looked on March 5, 1936." The meaning of the term "as it looked" becomes clarified if we replace the impression form by the similarity disjunction. This disjunction in the shorter form would read: "An object of the class of things similar to a flashlight, on March 5, 1936." The so-called "descriptions of impressions" occurring in usual basic statements are permissible only if the hypothesis of the constancy of the perceptual function is valid.

We know, however, that it is not always valid. There are well-known exceptions: putting our hand into a pot of water of a certain definite temperature, we may sometimes sense the water as warm, sometimes as cold, according as we have immediately before put our hand into colder or warmer water. In this case the water always shows the same objective relations to other physical bodies, expressed by the constant registering of the thermometer; but we sense it differently. Thus the perceptual function here is not constant. The case is different from the foregoing example (which we constructed artificially) in so far as the perceptual function is not a variable dependent on time directly but on the nature of the physical objects perceived immediately before. So we have to add not a time index but a remark concerning the objects previously perceived: we have to say, for example, "the feeling of hot water as it occurs after touching cold water." Other examples of this kind occur in optical sensations; the sensed color of a surface may depend on the color of the surrounding surface. In this case it is the spatially adjacent sensation and not the temporally adjacent one which is to be named in the exact description. Psychology has pointed out a number of similar cases, and we take notice of them in our observational technique. Setting aside these exceptional cases, we keep in general to the hypothesis of the constancy of the perceptual function.

This hypothesis, therefore, introduces a further element of uncertainty into basic statements in the wider sense. For it is obvious that, practically speaking, we can control this hypothesis only in certain cases and extend its validity from these by inductive inferences. If we add this to the foregoing results concerning the reliability of memory, we find that basic statements in the wider sense are by no means absolutely certain.

Our investigation thus confirms our idea that, if there are absolutely certain statements at all, these can only be basic statements in the narrower sense. The question remains whether statements of this kind may be absolutely certain or not.

The answer to this question can now be given. It reads that, even if there are such statements, it will never be possible to formulate them. Every formulation occupies a stretch of time, and during this time there may occur certain changes of the kind already indicated. We imagined, in our discussion of the reliability of memory and of the constancy of the perceptual function, a rather slow change of conditions, which furnishes observable differences only from day to day; but we cannot exclude the possibility that there is, or will be, a much quicker change, in which minutes or seconds take the place of the days in our examples. Human forms of speech cannot cope with such possibilities. Our basic statements in the narrower sense are, strictly speaking, basic statements in the wider sense in which the involved time interval is of short duration. Consequently there is only an approximation to basic statements in the narrower sense; and this implies that there is in any utterable proposition only an approximation to absolute certainty. Absolute certainty is a limit which we shall never reach.

We may be glad if there is at least an unlimited approximation, i.e., if it is possible to increase the certainty to any desired degree of probability, less by a small difference ϵ than certainty. There is, however, no proof that even this is so. Quantum mechanics showed that this unlimited approximation is not valid for predictions concerning future events; it may be that the same restriction holds for statements concerning the immediate present. However, this is of no important practical bearing because all statements which we can construct in practice are statements for which a remnant of uncertainty persists.

§ 22. Weight as the sole predicate of propositions

Our inquiry concerning impression statements has far-reaching consequences for the theory of truth.

Throughout the first chapter we entertained the presupposition that propositions about concrete physical facts, which we called observation propositions, are absolutely verifiable. A more precise analysis showed that this conception is untenable, that even for such statements only a weight can be determined. With the object of obtaining more reliable statements, we then introduced impression propositions; throughout the second chapter we upheld the supposition that at least these propositions are capable of absolute verification. We have discovered now that even this is not tenable, that impression propositions also can only be judged by the category of weight. Thus there are left no propositions at all which can be absolutely verified. The predicate of truth-value of a proposition, therefore, is a mere fictive quality; its place is in an ideal world of science only, whereas actual science cannot make use of it. Actual science instead employs throughout the predicate of weight. We have shown, in the first place, that this predicate takes the place of the truth-value in all cases in which the latter cannot be determined; so we introduced it for propositions about the future, so long as their events are not yet realized, and for indirect propositions, which remain unverified for all time. We see now that all propositions are, strictly speaking, of the latter type; that all propositions are indirect propositions and never exactly verifiable. So the predicate of weight has entirely superseded the predicate of truth-value and remains our only measure for judging propositions.

If we, nevertheless, speak of the truth-value of a proposition, this is only a schematization. We regard a high weight as equivalent to truth, and a low weight as equivalent to falsehood; the intermediate domain is called "indeterminate." The conception of science as a system of true propositions is therefore nothing but a schematization. For many purposes this conception may be a sufficient approximation; but, for an exact epistemological inquiry, this conception cannot furnish a satisfactory basis. An approximation is permissible always within a certain domain of application only, whereas outside these boundaries it leads to grave incongruity with the factual situation. The same holds for the schematized conception of science as a system of true propositions. In the hands of careful and not too consistent philosophers, it has not done much mischief; it has led instead to some unanswerable questions which have been modestly put outside the domain of solvable problems. But in the hands of pretentious and consistent logicians this schematized conception has produced serious misunderstandings of science and has led to grave distortions in the interpretation of scientific methods. In case of discrepancies between the constructed epistemological system and actual science the full weight of deductive method has outbalanced the unprejudiced view of the factual situation; instead of the deductive method being turned backward to a revision of the presupposed structure of science, this schematized structure has been abused as a support for a radical misinterpretation of the very nature of science.

This description seems to me to apply to the positivistic theory of meaning which makes meaning dependent on verifiability. So long as the demand of verifiability is not overstrained, that is, so long as a highly probable proposition is considered as true, this theory is a useful approximation; the greater part of scientific propositions can be retained as meaningful, future propositions and all kinds of indirect sentences included. But with the introduction of higher pretensions into the methods of analysis, a great number of the propositions of science are pointed out as unverifiable; the positivistic theory of meaning, then, expels these propositions from the domain of meaning and substitutes for them other sentences which, for any unprejudiced eye, cannot perform the functions of the condemned propositions. This procedure is carried through with more or less consistency; but none of its representatives has as yet had the courage to carry his principle through to its ultimate consequence and to admit that there are no meaningful sentences at all left in science.

The probability theory of meaning is free from such a dogmatism. If it admits verifiability in the sense of an approximation, it does not fail to recognize that even an approximate verification is possible for a group of sentences only and that in general the predicate of weight cannot be dispensed with. Thus the theory of meaning is constructed in a form wide enough to include as meaningful both verifiable propositions and propositions for which only a weight is determinable. When at last it is pointed out that absolute verification is a fiction never realized in practical science, this theory of meaning is not shaken; it is able to furnish the form of a generalized theory of meaning in which weight is the only predicate on which meaning is based. In this way a more general verifiability theory of meaning has been constructed in which verification is to denote only the determination of a degree of probability.

It is of some interest to survey, from this point of view, the train of our ideas. Our investigation started with the supposition that there are three predicates of propositions: meaning, truth-value, and predictional value. Applying the positivistic theory of meaning, we found that the predicate of meaning can be reduced to the predicate of truth-value; but expanding these considerations to indirect propositions, we discovered that this reduction furnished a too narrow concept of meaning and that we had to add the predicate of predictional value in order to obtain a wider basis for meaning. Verifiability in the wider sense, including the determinability of a predictional value, or weight—this was the quality upon which we made meaning dependent. Our last inquiry into the nature of impressions showed, however, that there are no propositions at all which are absolutely verifiable. It is in all cases the predicate of predictional value alone on which meaning is based. In this way the triplet of predicates, meaning, truth-value, and predictional value, has been reduced to one of these terms, to predictional value or weight. The concept of truth appears as an idealization of a weight of high degree, and the concept of meaning is the quality of being accessible to the determination of a weight. What we introduced as a bridge from the known to the unknown turns out to be the only measure of scientific thinking; the bridging principle has absorbed the other members of the triplet of predicates of propositions.

This result is in strong contrast to certain ideas which have been developed in defense of the truth theory of meaning. It has been argued that predictional value concerns only our subjective expectation and that it cannot furnish a basis for the definition of meaning; inversely, it is said, a predictional value presupposes meaning in the sense of absolute verifiability because we can expect only events which later on can be judged as having happened or not having happened. This objection is an example of the erroneous consequences to which the schematized conception of science may lead. It mistakes the fact that the so-called verification of the event, after its happening or not happening, is nothing but another determination of a weight, with the only difference that this weight is of a higher degree and can be approximately identified with truth. We pointed this out in the example of the cubical world, showing that a direct view through the walls could not absolutely convince us that there are birds outside but would only furnish us some new physical objects, the nature and localization of which would have to be found out by means of probability inferences. It is true that these inferences furnish a higher degree of probability for the hypothesis of the birds than could be obtained before. But this is all that can be maintained; there is no absolute verification. It is therefore not true that probability inferences can refer only to facts which are accessible to direct verification by other methods. The argument on which the objection is based would read in a precise formulation:

We can only expect, with any degree of predictional value, events which later on will obtain a higher predictional value. In this form, however, the lack of cogency is obvious.

The probability theory of meaning cannot be reduced to the truth theory of meaning; on the contrary, the latter must be conceived as a schematized form of the former, valid only in the sense of an approximation.

If, from this point of view, we take up the question of the positivistic construction of the world, we find that the introduction of the impression basis does not free us from probability statements, not even at the very basis itself. It is not only the inferences from the basis to external things which have a probability character; the same is valid for every statement concerning basic facts. This is the last blow against the positivistic theory, shaking even the last remnant of absolutism still left to it after the rejection of its wider pretensions. The psychological origin of this theory was the tendency to restore absolute certainty to all statements about the world; if statements about impressions were absolutely certain, and if statements about physical things were nothing but equivalent transformations of impression statements, this aim would be reached. We found in the preceding chapter that the second part of this theory is not tenable, that the relations between impressions and physical facts are probability relations, and that the certainty of the basis cannot be transferred to our knowledge of external objects. In the present chapter we found that a similar fate attends the basis itself in the light of a precise examination. There is no certainty at all remaining—all that we know can be maintained with probability only. There is no Archimedean point of absolute certainty left to which to attach our knowledge of the world; all we have is an elastic net of probability connections floating in open space.

CHAPTER IV

THE PROJECTIVE CONSTRUCTION OF THE WORLD ON THE CONCRETA BASIS

CHAPTER IV

THE PROJECTIVE CONSTRUCTION OF THE WORLD ON THE CONCRETA BASIS

§ 23. The grammar of the word "existence"

Our inquiry into the nature of impressions led us to the conclusion that impressions are not observed but only inferred. We said that the things directly observed are the concrete things of daily life and that it is an inference which leads us from them to the existence of impressions. The basis of the epistemological construction, therefore, is the world of concrete objects; from this sphere inferences lead to more complex physical objects, on the one hand, and to impressions, on the other.

It will be our task to analyze this process, to develop the whole construction of the world on the concreta basis—the result forms what is usually called our picture of the world. The analysis of this construction will furnish us a theory of existence which relates our results concerning the probability character of the combining relations to the discovery that it is the sphere of concrete objects, not of impressions, which should be taken as a basis for the rational reconstruction of the world.

Before entering into this analysis, however, we must make a preliminary remark concerning the term "existence." Language expresses this concept by the term "there is." If we ask for the meaning of this term, we must begin with an inquiry into the rules according to which the words "there is" are used. That is to say, we want to learn the grammar of the term; without knowing this grammar we should not be able to employ the term in an understandable way.

Entering into this inquiry, we must note first that the words "there is" do not always have the meaning of existence. If we ask "Where is William?" and receive the answer, "There is William," this "there is" expresses a spatial determination; we do not want to emphasize that William "exists" but that he is at the place denoted by "there." The meaning of existence is expressed in another kind of phrase. We say, for example, "There is a bird as tall as a horse"; the "there is" here does not indicate a spatial determination but that such a bird exists. This is obvious if we compare the last phrase with the phrase, "There is an ostrich," spoken, say, before the cage in a zoo; in this phrase "there" is a spatial determination, as in the first example. Let us consider the construction of a phrase containing the existential "there is."

The essential feature of such phrases is that they contain the term "there is" or "there exists" not as applied to an individual but in the context of a description. A description is a combination of words, the sense of each of which is already determined, but which defines, in combination, a new term. We can ask then whether there exists a corresponding thing. This is a reasonable question because we cannot infer from the description that such a thing exists; this is not possible even in case the existence of things corresponding to the constituents of the definition is guaranteed. If we know that there exist a mammal and also an animal with a trunk instead of a nose, we do not yet know that there exists also a mammal with a trunk. This is why language applies here the concept of existence and formulates the sentence: "There is a mammal with a trunk." This proposition informs us of something new; its truth is confirmed when we see an elephant. What is stated, however, is not the existence of this single elephant but of a thing corresponding to the given description. An existential proposition always concerns the existence of the specified, not of an individual.

Logistic expresses this idea by the prescription that an existence sentence is always to contain an operator together with a bound variable:

$$(\exists x)f(x) \tag{1}$$

which formula reads, "There exists an x such that $f(x)$ is true." We never write $(\exists a)$, where a is an individual; i.e., we do not say, "This elephant exists." Such a statement would be meaningless. If we have the feeling that perhaps this statement means something, this is because we take the word "elephant" not in the sense of an individual sign but in the sense of a description. A manual of zoölogy contains a description of an elephant; if we point to an elephant and say, "This elephant exists," this may mean "This thing exists as an elephant," or more briefly, "This thing is an elephant." It is obvious that the word "elephant" in all these phrases is a description. If we were to point to the elephant and say, "This lion exists," our assertion would be false not because the elephant does not exist but because it is not a lion. If the phrase "This elephant exists" is accepted as meaningful, the word "elephant" must therefore be a description, and our phrase must be interpreted as meaning "There exists an elephant in the direction in which I point," or simply: "This thing is an elephant."

The last phrase does not contain the concept "existence," for the word "is" in this case is the copula and not the existential "is." So the form of the last phrase is $f(a)$, that is, a certain predicate f (being an elephant) is predicated of the argument a. We see that a statement of such a kind may be used as a substantiation of an existence proposition. If the thing *a* is an elephant, we are correct in saying "There is an elephant." In the last form, the "is" is the existential sign, and "elephant" is a description. Logistic expresses this relation by the formula

$$f(a) \supset (\exists x) f(x) \tag{2}$$

We may say: The thing *a* confers existence on a corresponding descriptum. This is the correct way of expressing the relation between things and the term "existence."

§ 24. The different kinds of existence

This point in grammar having been determined, we shall now proceed to a further analysis of the concept of existence. The next thing to be noted is that the concept of existence divides into different subconcepts which must now be explained.

Imagine we are taking a walk at dusk through a lonely moor; we see before us at some distance a man in the road. He is a strange little man, wearing a caftan, and carrying a bag on his shoulder. In spite of a certain feeling of uneasiness we do not doubt the man's reality. Coming nearer, we see that he does not walk; he stands and waves his hand. We advance farther and discover that it is not a man that we see there but a juniper bush, a branch of which is moved by the wind.

What has happened in this case, logically speaking? First, there was a man and, afterward, a juniper bush. We know, now, that the juniper bush is the "real" thing and that the man was an "apparent" thing only; but this man had an existence in a certain sense. We may even go back several steps and "produce" the man once more, in spite of knowing about the illusion. The juniper bush then does not cease to exist—that we know—but we do not see the bush; we see the thing like a man and not like a bush. We shall say that both the man and the bush have *immediate existence* at the moments we see them. In spite of this common quality there is a difference as to their existence: the immediate existence of the man is a *subjective existence* only, whereas that of the bush is an *objective existence.* We must add that the objective existence of the bush may even persist when its immediate existence has ceased whereas the subjective existence of the man is bound to the duration of the immediate existence.

It follows that the three new terms introduced denote partially overlapping subclasses of the existence concept.

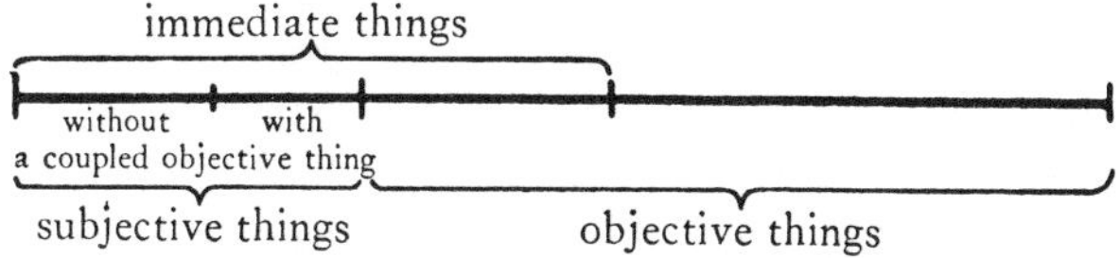

FIG. 4.—The different kinds of existence

Existence is divided into subjective and objective existence; the domain of immediate existence, however, includes all subjective existence and, in addition, a part of the domain of objective existence. It is this domain of immediate existence to which our epistemological interest will be particularly directed.

According to our new notation, we shall also apply the terms introduced to things directly. We shall speak of subjective and objective things and of immediate things. This mode of speech will facilitate our investigations. Figure 4 may illustrate our classification.

The subjective things involve a further subdivision. The subjective thing of our example, the man with the caftan, stands in a certain relation to the objective thing, the bush; we should not see the man if there were no bush, and we see that the man is altered if the bush is altered. If the branch of the bush is moved, the man waves his hand. We shall say in such a case that the subjective thing is *coupled* to a certain objective thing. Our observation of the subjective thing, in this case, is bound to an observation of an objective thing in the physical sense, i.e., in the sense that light rays coming from the bush enter our eyes; we do not observe, however, the bush as a bush, but as a man. There is then no immediate bush; what exists instead is a subjective man.

A very instructive case of this type is that of the cinema. The immediate things we see there are of a very suggestive character; though we know they are subjective things only, we cannot withstand their intuitiveness, their persuasiveness, and are seized by them in such a way that emotions of pain, affliction, joy, tenseness, and sympathy are aroused as though the subjective things were objective. The co-ordinated objective thing is here the screen as a sheet of cloth, or a whitened wall covered with dark and bright patches. The objective and subjective things are coupled; a movement of the patches on the screen produces a movement of the subjective things. In this case, however, the subjective and objective things do not always occupy the same place in space. The subjective things have a certain spatial depth and therefore cannot be localized on the two-dimensional screen. They may even be very far off; such is the case in a view of distant mountains which by the perspective of the picture may subjectively appear at a distance of some miles.

There are, however, cases in which there is no objective thing co-ordinated with the subjective thing. Such is the case of dreams. The subjective things here are also very suggestive and are not associated (as in the cinema) with a knowledge about their merely subjective character. In this case, however, there is no coupled objective thing at all. That is to say, when I dream that my friend stands before me, there may be objective things standing just at the place where my friend is localized; but they are not coupled with my friend in the sense defined (certain movements of these other things do not produce corresponding movements of my friend).

One might be tempted to construe another difference between the cinema and the dream by pointing to the fact that the subjective things in the cinema correspond to some objective things actualized at an earlier time, namely, to the movements of the actors during the taking of the film, whereas there is no such correspondence for the dream. This difference, however, is not relevant for our considerations. We do not call the correspondence between the cinema pictures and the actors a coupling; if we speak of an *existential coupling*, this coupling is to concern states of things existing at the same time. It is this concept of existential coupling on which our subdivision of subjective things is based.

The subjective things both of the cinema and of the dream are immediate things; in this respect they do not differ from such objective immediate things as the physical things of our daily environment. The separation of immediate things into subjective and objective cannot be performed on the basis of immediate intuition; their intuitiveness is their common feature, and we must apply other methods to separate them, methods of which we shall speak later. What is meant by immediate intuitiveness is not to be defined; we may regard immediate existence as a concept known to everybody. If someone does not understand us, we put him into a certain situation and pronounce the term, thus accustoming him to the association of the term and the situation seen by him. We make use here of the same method as employed for the definition of special empirical concepts. If a child asks us, "What is a knife?" we take a knife and show it to the child. It was in this manner that we first learned the sense of words, that is, the correspondence of words to things. We previously presented this idea (§ 5) by imagining a collection of specimens in which everything bears a label with its name on it. We pointed out also that qualities such as "possession" or "being larger than" are to be demonstrated in the collection of specimens; there may be two poles of different size, marked as "pole *a*" and "pole *b*," and a label inscribed, "Pole *a* is larger than pole *b*." In the same way, the concept of immediate existence could be presented. After our visitor has passed before many cages, each with a label bearing the name of the animal, he is led to a large cage in which many different animals are moving about. "There is an elephant among these animals" may be written on a label before this cage. The term "there is" occurring here stands for our concept of immediate existence. If it is introduced in the form described, it is simultaneously shown that, as we remarked in our grammatical excursion, existence always concerns a description, that the words "there is" denote the existence of the specified among other things. That this term is not limited to objective things but applies to subjective things as well may be pointed out by the fact that a dreamed collection of specimens of the arrangement described would suffice for the explanation of the intuitive "there is" as well as the real one.

After the determination of the concept of immediate existence we must turn to the concept of objective existence. This concept is of a type entirely different from the first. Objective existence is not an intuitive quality; it must be determined by relations which are attached to the concept of immediate existence. That is to say, objective existence is a determinate logical function of subjective existence.

To carry through this determination, we have to reconstruct the methods by which the distinction of immediate and objective existence is performed in practice. In the pursuit of this plan, we turn next to the task of expounding the logical construction of the system of knowledge.

§ 25. The projective construction of the world

The original world is the world of immediately existing things. It is the world of concrete objects around us, entering into our knowledge without any intellectual operations being performed by us. It is a world where there is no difference between waking and dreaming; in which everything exists exactly in the form in which it is observed.

The word "original," with which we characterize this world, has three significations. First, it means that this is the world which *historically* is first, standing at the beginning of the long road which has been traveled by mankind from its primitive stages to the complicated state of intellectual culture of our day. Second, it means that this is the world at the beginning of the *individual* mental development of any human being, i.e., the world of early childhood. Third, it means that this is the *psychologically* first world; by this term we mean that this is the world which presents itself immediately, which is actually not constructed by inferences but is the basis of all inferences actually performed by us.

There is a theory that there remains a question as to the *logically* first basis, i.e., a basis which must be chosen for logical reasons as the ground of all inference if we want to give the rational reconstruction of the world. This idea seems to me untenable. Logic does not distinguish one basis as the necessary one; logical inferences may be attached to any basis, and what is a basis for one logical system may become a deduced result for another. This logical arbitrariness of the epistemological basis has been justly pointed out by Carnap.[1] If we want to mark one basis as the "original" one, this question may only concern that basis which corresponds best to the actual performance of knowledge; we may ask for the best adapted form of the rational reconstruction. This leads to the three senses of the word "original" as distinguished, according as we want to adapt the rational reconstruction to the historical course of knowledge, or to the course of the individual acquisition of knowledge in the development from childhood to manhood, or to the course of operations in which knowledge is actually performed at every moment in which we want to know something new. These three kinds of basis are perhaps not identical but they are similar and surely rather remote from the "simplest" basis such as logicians would like to assume. Seen from the viewpoint of a neatly ordered system, in the logical sense, the actual basis is on a rather complicated middle level. This is especially obvious if we consider the basis in the third sense. The act of acquiring knowledge reveals its implicit basis whenever doubts of the physical world occur, as, for instance, at the moment of awakening, or at times of high nervous tension. We go back then to the immediately existing objects, to the concreta, as the most reliable facts. This *return to the basis of immediate existence* points out that it is the world of the concreta which forms the actual psychological basis.

Let us consider this original world and the ways in which we emancipate ourselves from it. Primitive people make no distinction between subjective and objective existence; they take as real what they observe, and they know no difference between dreams and wakefulness. Explorers re-

[1] R. Carnap, *Der logische Aufbau der Welt* (Berlin and Leipzig, 1928), p. 83.

late strange stories about the interconnection of dreamed and real facts among primitive races. A man who dreams that a certain woman makes a declaration of love to him may take this as a real offer; a man who dreams that another man wounded him, or some member of his family, may try to kill this man.[2] Observations of children in the days of early childhood furnish analogous results; we know that there are children who relate, without any consciousness of lying, things which never happened, as if they were observed by them—thus revealing that they do not always differentiate between subjective and objective existence. We see that it is not only the difference between dreaming and being awake which is in question here. There are many things seen while awake which afterward turn out to be of merely subjective existence. To this class belong optical illusions like the image seen in a mirror, taken originally as a material thing behind the mirror, or the appearance of the bent stick produced by a straight one put into clear water. Originally the world is full of illusions of this kind. Historically speaking, it was a long time before mankind learned to distinguish between subjective and objective existence, a distinction obtained by means of intellectual processes but not directly furnished by observation.

The logical way in which this distinction is made is as follows. We begin with the presupposition that all things which we observe exist; that is, with the presupposition that immediate existence is equivalent to objective existence. We contrive then to construct a net of combining relations between the things; we call these physical laws. They are relations of the type, "If there is one thing, there is another thing also." If the other thing is not observed, it is easy—in this primitive state—to alter certain conditions and thus observe it. The primitive man sees that

[2] Cf. Lévy-Brühl, *La Mentalité primitive* (Paris, 1922), p. 102.

there are certain traces in the sand and infers that there is a bear; he then goes into the woods and sees the bear. Thus we succeed in constructing inferences on the basis of observed relations which lead to foreseeing future events.

In performing inferences of such a kind, however, we discover that we are not always successful. The analysis of this fact leads to the discovery of the dream world. The primitive man may have "seen" a bear before his cave, but afterward he finds neither traces in the sand nor the animal itself in the wood. Analogous inferences show the unreality of our own dream world, which is occupied with subjects other than those which concern the primitive man. But it is not only the difference between dreaming and being awake which is established in this way; it is the totality of all other corrections of our immediate world as well. When we try to touch a thing seen in a mirror, at the place where it is seen, we touch nothing; this is the way in which we discover the "virtual" character of the image in the mirror, a method actually performed by children, and even monkeys, when we put a mirror before their eyes. The laws of nature involve contradictions if we consider the whole immediate world as real—this is the reason that the distinction between the objective and the subjective world is introduced.

The method described is a typical statistical method. It starts with the presupposition that all things are real, and arrives at the result that some of them are not real. There is no contradiction in this method, though it cannot be replaced by another which needs no presupposition to be refuted later on. The presupposition is the identification of immediate and objective existence; the result is the division of the domain of immediate existence into a subjective and an objective part. We may say that the character of immediate existence entitles us to assume a thing as having the character of objective existence so long as no contradiction arises.

The statistical character of the method is expressed in the acknowledgment of the superiority of the greater number. The objects of the waking-world are more numerous than those of the dream world; therefore the waking-world is conceived as the "normal" world, the dream world, on the contrary, as the exception. There is a kind of democracy in our subjective world, and the dream world is outvoted. However, this is not the essential point; there is another quality of the waking-world which distinguishes it from the dream world.

This second point is a statistical matter also but of another type. We said that we construct predictions by making use of the laws of nature. If we now count the success ratio of the predictions, we find that we have arrived at a much better success ratio if we have put the things of the dream apart and do not use them as basis for predictions. This is illustrated in the case of the man who dreams of a bear in his cave but does not observe afterward the traces in the sand, or the bear in the wood. Even if the world of dreams were quantitatively superior to that of wakefulness, the latter would be denoted as superior by this quality of admitting predictions. We cannot construct laws dealing with the things dreamed and furnishing predictions which are confirmed within the dream, or within another dream.

There is a third point of a statistical character which is in favor of the waking-world. It is possible to combine both worlds into a single one if we leave the things of waking as they are but interpret the things of the dream in a way quite different from their immediate appearance. That is to say, if we interpret the things we dream as merely subjective, but as due to internal processes in our body which have objective existence, we arrive at a single world in which prediction is possible, even when the dream world is included. We can, on the one hand, foresee the dream world to a certain degree; we know that after a certain exciting experience we shall dream of it, we know that after taking a soporific the dream world is suppressed, etc. We can, on the other hand, use the contents of dreams for predictions concerning the world of waking; this is a rather modern discovery owing to Freud's psychoanalysis and applied in psychical cures. This is the epistemological significance of psychoanalysis; it showed for the first time how to construct a causal connection between the two worlds of waking and dreaming. The objects of the dream in this context are not considered as physical objects but as pseudo-objects indicating certain states of the nervous system of the human body. This third point is statistical, like the second, because it cannot furnish an absolute decision in favor of the world of waking; it furnishes only a statistical decision because the laws obtained are probability laws only, i.e., valid in the greater ratio of events.

From the statistical character of the inferences occurring here it is obvious that we never obtain an absolute certainty about objective existence. This corresponds to the result of the preceding chapters. A statement that a certain thing objectively exists is never absolutely certain, be it even one of the simple and concrete things of daily life. But the degree of weight obtained in such a case is, of course, rather high.

It is not always necessary to carry through the whole statistical method in order to discover the merely subjective character of certain objects. Basing our inference on many former experiences, we learn to discern subjective and objective things immediately. As for the dream, we perform this distinction immediately after awaking, without needing further experience; in other cases, the appearance of the object is accompanied by a knowledge of its merely subjective character. This is the case of so-called images which we produce intentionally, or which are raised in the context of other experiences, by association, etc. To explain this, we might speak of a *scale of gradation* of the immediate existence character; representations have a rather feeble existence character if they are produced intentionally but may acquire a stronger existence character if they arise spontaneously. Objects appearing with a feeble existence character are not regarded as real, i.e., we know immediately that chains of inferences attached to these objects would lead to contradictions, and we need not carry through the statistical method. This renunciation of control is perhaps, psychologically speaking, a result of former experiences in early childhood; in any case, it can be logically conceived as such. This means that in the rational reconstruction of knowledge we might start with the presupposition that all objects, the representations included, are real, and prove then by our statistical methods that the representations are not real. Certainly this procedure is used by us every time we are in doubt as to the reality of an observed object. There are sensations with a very feeble degree of existence character, such as sensations outside the field of concentration, as in the case of optical sensations within the peripheral optical field; to clarify their reality, we control them by inferences leading to sensations of a stronger existence character. Thus we turn our eyes in such a way that the supposed thing enters into the central optical field; if it is observed, then we infer that the object formerly seen was real. This is an example of what we called the return to the basis of immediate existence; in a case when we are uncertain about objective existence, we go back to the presupposition that what has immediate existence also has objective existence, and control this presupposition by the statistical method.

Although we may, in cases such as those described, interpret a low degree of existence character as indicating the subjective character of the object, we must not invert this relation: a high degree of existence character does not necessarily involve the objective character of the thing. There are things of a high degree of existence character which are only subjective; their subjectivity may even be known to us without any enfeebling of the existence character being involved. Of this kind are the things seen in a cinema. We know from the whole situation, from the surrounding interior of the theater, etc., that these things have no objective existence; but their immediate existence is of so high a degree that we submit to the suggestion of their reality and forget, for a while, their merely subjective existence. In this case the knowledge of the unreality of the seen objects is certainly psychologically acquired by former experiences. Small children when taken into a cinema take the pictures for real beings and may be afraid of the terrible beasts and men they see there.

However, the great majority of the things of daily life, the concreta, are, for us, real beyond any doubt. This is because they have stood up to every test ever applied. We are entitled to identify their immediate existence, being of so high a degree, with objective existence. This is the reason that these things are so concrete, so indubitable, so solid in their intuitive reality. It is the combination of immediate and objective existence character which is the essential feature of concreta.

The concreta form the basis of inferences which lead to the existence of other things. That is to say, the inferences leading from immediate to objective existence are for concreta skipped in practice; once the existence of concreta has been ascertained, inferences from them lead to other things of a less immediate character.

There are, first, inferences to other concreta. The domain of concreta accessible to direct observation is restricted, on practical grounds, and for every person in a different way; our personal situation in life allows us to enter into direct contact with only a restricted number of things. There are other continents, foreign people, unseen machines, which we infer from our surrounding concreta, without the possibility of observing them directly. But this is only a technical impossibility, and we call these things concreta also. Though they never had immediate existence for us, they might obtain it; we provide a substitute by looking at pictures, i.e., by bringing similar things into immediate existence. The inferences leading to these things are probability inferences; we are never absolutely sure whether these other concreta actually exist. But this uncertainty is not relevant; it does not render our situation appreciably less secure, as even the existence of accessible concreta is not absolutely certain.

Second, there are inferences to abstracta. These inferences are, as we pointed out in § 11, equivalences, not probability inferences. Consequently, the existence of abstracta is reducible to the existence of concreta. There is, therefore, no problem of their objective existence; their status depends on a convention. As for immediate existence, it may be taken as a definition of abstracta that they have no immediate existence. Actually the determination of abstracta is somewhat arbitrary, so that the term "abstract" itself is rather vague. There are many cases in which we are undecided whether a term is an abstractum or a concretum (cf. § 11). The process of forming abstracta may be continued to the formation of abstracta of higher levels, the elements of which are already abstracta. Thus abstraction involves a direction; on the higher levels the decision as to the abstract character of the terms becomes more determinate.

Third, there are inferences to other things which are not abstracta, but which cannot become concreta either, since, for physical reasons, their becoming immediately existent is precluded. Of this kind are things such as electricity, radio waves, atoms, or many invisible gases. The existence of these things is not reducible to the existence of concreta because they are inferred by probability inferences from concreta. Let us introduce the term *illata* for these things, i.e., "inferred things."[3] We see that the old disjunction of concreta and abstracta is incomplete; a third term is needed to denote things which are neither concrete—capable of immediate existence—nor abstract—reducible to concreta. The relation of the illata to the concreta is a projection in the sense indicated in § 13. The illata have, therefore, an existence of their own, as the birds for the people of the cubical world, although they are not accessible to direct observation, that is, to immediate existence.

If it is questioned whether the illata are logically different from the abstracta, i.e., if it is maintained that the illata are reducible to the concreta, we must answer with the arguments developed in the discussion of the cubical world (§ 14). Our observations of concrete things confer a certain probability on the existence of the illata—nothing more. It is not possible to enlarge the class of the considered concreta in such a way that statements about this class are equivalent to a statement about the illatum. The equivalence maintained by positivists is due to the neglect of the probability character of the inferences. The atoms

[3] We use the participle *illatum* of the Latin *infero*, to denote this kind of thing.

have been discovered by the physicists in a way analogous to the discovery of the birds in the cubical world. Certain observed relations between macroscopic bodies—such as expressed in Dalton's law of multiple proportions—made it very probable that all bodies are built up of very small particles, though these particles could not be directly observed; this was the first substantiation given by the physicists to the theory of atoms. Mach, from his positivistic standpoint, declared that the concept "atom" was nothing but an abbreviation for the relations observed between macroscopic bodies; in our language: Mach declared that the atom is a reducible complex of concreta as internal elements. Boltzmann, one of the leading investigators in the domain of atomism, opposed Mach's "dogmatism" and defended the independent existence of atoms; he compared the hypothesis of atoms to the hypothesis of the stars as being enormous bodies at enormous distances—a hypothesis, he said, inferred only from "scanty optical sensations."[4] To this hypothesis, he continued, it could also be objected that it constructs "a whole world of imagined things in addition to the world of our sensations"; but in this case nobody doubts their reality. Boltzmann's argument in our terminology would read that there are probability inferences to the existence of atoms, that the atoms are a projective complex of concreta, and that it is no objection against the independent reality of the atoms if a "direct verification," i.e., the determination of a higher weight is physically impossible. Later developments have decided in favor of Boltzmann's opinion; effects have been discovered experimentally which are comparable to a penetration of the walls of the cubical world as described by us. These are the famous discoveries which show individual

[4] "Von spärlichen Gesichtswahrnehmungen" (L. Boltzmann, *Vorlesungen über Gastheorie* [Leipzig, 1895], p. 6).

effects of a single atom, or electron, like the Wilson tracks of alpha and beta particles. It is true that they do not show the individual atom in the same way that we see a tennis ball; but they increase the weight of the hypothesis to such a degree that no practical doubt remains.

It may be answered that it is unavoidable that our direct observations concern macroscopic objects, that the objects seen in the verification of the atomic hypothesis, such as the Wilson tracks, are macroscopic objects also, and that therefore the meaning of the concept "atom" can never be more than a statement about concreta. To this we have the two answers developed in the example of the cubical world. The first answer is that such an epistemological theory presupposes physical truth meaning, and that with such a meaning the existence even of the concreta cannot be maintained as meaningful; that physical probability meaning, however, allows us to speak meaningfully of atoms as independent entities. The second answer is that with logical meaning the existence of the atoms is directly verifiable if we confine ourselves to practical verifiability. It is, physically speaking, an accidental matter that we cannot see atoms, owing to our being of larger dimensions. It is not logically impossible that some day we shall learn to diminish our size to submicroscopic dimensions and to observe atoms directly. We refer for these reflections to §§ 6 and 14.

The latter argument, to give it a less abstruse form, may be interpreted in the following way. The human body so far as its size is concerned happens to be situated in a certain range of medium physical sizes; it possesses sense organs reacting to certain physical processes only, yielding impressions only of things of medium size and medium intensity. By this physical place of our bodies in the world, the class of our concreta is determined. Smaller beings or beings of other sense organs would directly observe what we must infer; men with eyes structurally different from ours would see radio waves as we see those of light and would not need to infer them from sounds or pictures. Larger beings would see directly as a whole what we must construe as abstracta; they might see our planetary system as a whole, as a celestial merry-go-round. The division into concreta, abstracta, and illata, is therefore not a matter of principle, but due only to our personal situation in the physical world. Consequently we should not make any distinction as to the existence of objects corresponding to these terms.

This is to say that the world of concreta is only the first step in our construction of the world. From this step, we construct the abstracta as reducible complexes, the illata as projective complexes. Abstracta and illata have as a common feature their inaccessibility to immediate existence; but, in respect to objective existence, their logical character is entirely different. The objective existence of abstracta is reducible to concreta, so that these are internal elements of abstracta. The objective existence of illata, however, is not reducible to concreta; these are external elements of illata as projective complexes.

It might be asked whether it is possible to introduce, instead of concreta, other basic elements which would be elements internal to all objects. This is the question of the atomic theory of physics. Modern physics has shown that electrons, positrons, protons, neutrons, and photons, are the basic elements out of which all things are built up in the form of reducible complexes. For this basis, however, not only abstracta and illata but concreta as well are reducible complexes. The logical character of this basis, as a basis of internal elements, provides a good illustration of the logically different character of the concreta basis (and of the impression basis as well). The latter is a basis of external elements, upon which the world is constructed by projection.

These reflections necessitate an additional remark. We called the atom a projective complex of concreta but, on the other hand, said that the atoms are internal elements of concreta, as reducible complexes. This seems to be a contradiction, but the paradox is resolved when we distinguish the physical relations between things from the way in which we discover them.

The relation of reducibility is an objective relation, but there are different ways of establishing it. The ways differ according to what is given as a starting-point. If the elements are given, together with the relations between them, the complex is constructed by definition; this is the way we construct the abstracta. In the case of the atoms, however, the complex is given, and the elements must be inferred. Since all observable qualities of the macroscopic bodies are only averages of qualities of the atoms, there are no strict inferences from the macroscopic bodies to the atom but only probability inferences; we have, therefore, no equivalence between statements about the macroscopic body and statements about the atoms but only a probability connection. The relation is consequently of the logical type of a projection. However, it is a projection somewhat different from that analyzed in the example of the birds and their shadows, as it leads to things which are the internal elements of the things from which the inference started. Let us speak here of an *internal projection*. It is a projection because it establishes a probability connection between propositions; but the propositions obtained maintain that there is a reducibility relation. Thus the occurrence of a reduction is in this case ascertained by probability inferences, not by definition. Consequently it is not absolutely certain that the maintained reducibility holds; in this case, the reducibility is an empirical result. The internal projection, has, in common with the external one, a probability character, but it differs with respect to the existential relations. As it leads to internal elements, the existential relations here correspond to those of reduction (cf. § 13), with the sole difference that the validity of these relations cannot be maintained with certainty.[5]

We said that abstracta and illata are not accessible to immediate existence; the limits, however, are not sharply demarcated and may even be shifted by psychological processes. We do not observe air in the sense that we observe water; we do not see the state as a political body in the sense in which we see a marching regiment of soldiers. We cannot "realize" them in the sense of representing them with the character of immediate existence. We try to fill up the concepts as much as possible with "intuitive sense," i.e., we imagine some of their characteristic features which have the character of immediate existence. We imagine the feeling of wind and the resistance felt in pumping a tire, to realize the meaning of "air"; we think of public buildings, of marching soldiers, of a trial, with the intention of attaching the feeling of existence to the word "state." The word "realize" characterizes this process by its linguistic origin; it means originally, "making real," and we understand the metamorphosis of the word when we interpret its secondary sense as "transferring immediate existence to a thing." In this linguistic transformation, the concept "real" and the concept "immediately existent" have been assumed to be identical.

[5] There are other examples of an internal projection in which both sides of the co-ordination are directly observable; e.g., the case of a leaf and its cells which are visible under the microscope. The fact that there is a reduction of the leaf to the cells is, as in the example of the atoms, an empirical result and not maintained with certainty.

The process described may be denoted as the acquisition of an intuitive character by abstracta and illata; it cannot be arbitrarily extended but is governed by psychological laws. Only to a certain degree may this process be extended. It may happen, on the other hand, that we lose a distinct knowledge about that which may be called "immediately existent." Familiarity as to the use of a concept may be taken as intuition. If the electrician believes that he has an intuition of electricity, in the sense he has of running water, his usage of words seems scarcely permissible. In such a case some sensible effects of electricity are taken as representing the intended thing; the concreteness of the representatives is confounded with that of the original. But such psychological processes happen frequently and may acquire a great deal of practical value; they show in any case that the boundary between immediate existence and objective existence is indeterminate. They show at the same time that the "feeling of existence" is not an essential quality of objective existence but only an associated attitude.

It may be added that the character of concreteness is not restricted to things of material existence but may be attached to things which, physically speaking, are only "processes." We see the waves of the sea move as concrete things, but we know that there is no material thing moving with them, that they are to be explained as phase relations between vertical motions of water particles. A musical melody for us is a very concrete object, although it consists, physically speaking, of relations between individual tones. The pressure of a heavy load on our back is felt as a concrete power. Even the spiritual power of a great personality may be felt by us as a concrete entity; the illustration in ancient pictures of spiritual power by a halo shows the material conception of this power in all its concreteness in archaic minds. The domain of concrete things is not restricted to things of a spatial character; it is not at all determined by the place of the things in the physical arrangement of the world, but by psychological conditions.

These considerations, detailing the difference between the subjective and the objective arrangement of the world, show us the one-sided character of the perspective in which we see the world from the standpoint of our middle-scale dimensions. We walk through the world as the spectator walks through a great factory: he does not see the details of machines and working operations, or the comprehensive connections between the different departments which determine the working processes on a large scale. He sees only the features which are of a scale commensurable with his observational capacities: machines, workingmen, motor trucks, offices. In the same way, we see the world in the scale of our sense capacities: we see houses, trees, men, tools, tables, solids, liquids, waves, fields, woods, and the whole covered by the vault of the heavens. This perspective, however, is not only one-sided; it is false, in a certain sense. Even the concreta, the things which we believe we see as they are, are objectively of shapes other than we see them. We see the polished surface of our table as a smooth plane; but we know that it is a network of atoms with interstices much larger than the mass particles, and the microscope already shows not the atoms but the fact that the apparent smoothness is not better than the "smoothness" of the peel of a shriveled apple. We see the iron stove before us as a model of rigidity, solidity, immovability; but we know that its particles perform a violent dance, and that it resembles a swarm of dancing gnats more than the picture of solidity we attribute to it. We see the moon as a silvery disk in the celestial vault, but we know it is an enormous ball suspended in open space. We hear the voice coming from the mouth of a singing girl as a soft and continuous tone, but we know that this sound is composed of hundreds of impacts a second bombarding our ears like a machine gun. The concreta as we see them have as much similarity to the objects as they are as the little man with the caftan seen in the moor has to the juniper bush, or as the lion seen in the cinema has to the dark and bright spots on the screen. We do not see the things, not even the concreta, as they objectively are but in a distorted form; we see a *substitute world*—not the world as it is, objectively speaking.

Using the terminology developed above, we should say that even the concreta are only subjective things, of the type to which an objective thing of different form is co-ordinated. These things are coupled, but they are not strictly speaking identical. If we compare this co-ordination to that of our former examples, the juniper bush seen as a man or the cinema, we may say that, in the case of concreta, the correspondence of the subjective and the objective thing is closer than in those examples; but there always remains a deviation. This is the reason that the separation of objective and subjective things, within the realm of immediate things (§ 24), involves an element of arbitrariness; it depends upon what degree of deviation is to be tolerated for an immediate thing which is to be called objective. There is only a difference of degree between immediate things such as those seen in a cinema and immediate things such as the concreta: our immediate world is, strictly speaking, subjective throughout; it is a substitute world in which we live.

This fact is due to a psychological phenomenon which is connected with the logical structure of the existence concept. We showed (§ 23) that existence is a quality not of individual things but of *descripta;* only if a thing is given by description can we ask whether it exists. The mechanism of sensation is organized in such a way that it cannot produce a sensation without superimposing upon it a certain description. We do not see things as amorphous but always as framed within a certain description. It is as though we looked at a Persian carpet: its pattern consists of colored designs arranged in a strange and complex regularity; we may conceive its forms in different ways, grouping different forms as a whole—but we cannot visualize it without some structure. In the same sense the objects of our sensations always have a "*Gestalt* character." They appear as if pressed into a certain conceptual frame; it is their being seen within this frame which we call immediate existence.

The description in whose frame we see things corresponds to the objective thing only to a certain extent. This fact finds its expression in the predictional qualities of the co-ordinated description. To every description belongs a domain of included predictions; the degree of correspondence is measured by the ratio of true predictions within this domain. We see once more that between the subjective things of the kind occurring in the cinema, and the immediate concreta there is only a difference of degree: the ratio of true predictions is greater in the case of the immediate concreta—this is the only difference. Neither is the ratio of true predictions equal to *one* in the case of the concreta, nor is it equal to *zero* in the case of the cinema; in this case, also, there are a number of true predictions —those restricted to changes in the optical sphere—included in the description. The descriptional frame in which we see the world is never more than a substitute for a completely true description and will express only certain more or less essential features of the physical object.

The psychological origin of this frame must be supposed to lie in certain simple intellectual operations belonging to the primitive state of mankind or even to higher animals. Primitive man adapted his way of seeing to the simple cases of physical objects around him and to what he knew about these objects. He knew, for instance, that the tree he saw might be touched, that another tree partially hidden by the first could be reached after a greater number of steps (i.e., was more distant), and that the same tree would be seen by him on the following day, in the same place. Although this was not consciously formulated knowledge, it was knowledge instinctively acquired and expressed in his actions; in our rational reconstruction we have to express this fact by saying that he learned to attach to every observed object a group of inferences leading to other objects to be observed in the future. This acquired knowledge influenced his way of seeing; he came to see objects in the frame of a certain description. It is this primitive transition from immediate to objective existence which determines the form in which we see the world today—which creates the substitute world within which we wander throughout our whole life. Our immediate world is the objective world of primitive man; we see the world through the eyes of our ancestors, or, better stated, we see it as interpreted according to the knowledge of our ancestors. This primitive knowledge furnishes the frame of description into which we automatically press things in seeing them.

We need not refer to modern physics to show the discrepancy between the immediate and the objective world. There are simple and well-known phenomena indicating this difference. The object seen in a mirror is localized at a place corresponding to the place at which the material object usually stands when it emits light rays into our eyes. The psychological phenomenon of localization is adapted to the simplest and most natural case of observation as is proved by this example. We cannot alter this optical localization, but we can learn at least to alter some motor associations, to perform some manual operations upon an object seen not directly but only in a mirror. We see the stick put into the water bent at the point of its entrance into the water—i.e., we see it corresponding to a description which would be objectively true if the same optical datum were to occur outside the water under ordinary conditions. We see the rails of the railway converging toward the horizon; this means that we see them in the form shorter rails would objectively have if they were to offer the same optical effect. The phenomenon of the convergence of parallels may be conceived as an undervaluation of distance in the dimension of depth; in cases of shorter depth the parallels are not seen as convergent, as when we regard the edges of a book placed before us on the table. Our optical mechanism for erecting the optical image in the spatial form is adapted to small distances only; for greater distances it furnishes a substitute which would fit for the case to which it is suitable, that is, in the case of shorter distances. When we go straight ahead by railway, we see the flat fields turning around a distant but indeterminate center. This phenomenon comes about because our eyes cannot otherwise account for the perspective displacements observed—a fixed point at a certain middle distance appears at rest because our eyes follow it, whereas the more distant points move with the train, and the nearer points toward it. When we move more slowly, in walking, this phenomenon does not occur; our eyes are then able to correct the displacement of perspectives which is qualitatively of the same type, and to interpret it as a movement of our own body. This is once more an example of our optical apparatus being adapted only to the simpler case but furnishing a substitute in the more difficult case.

The substitute world around us is a product of the physical and historical conditions into which we are placed—a product of our situation in the middle of the physical world and at the end of a long historical development from primitive life to our present state. Analogous conditions are still at work and influence our vision. The social milieu into which we are caught adds pressure to the stronger influence of the physical and historical milieu. Our modern eyes, familiar with rectangular houses and steel constructions, see the richer forms of nature within the frame of our architectural style; modern drawings, in comparison with ancient drawings, betray this influence.[6] Instead of freeing our immediate world from the influence of our milieu, we adapt it to another milieu.

Must we renounce the possibility of ever obtaining a true picture of the world? I think not. Intellectual operations have shown us the way to overcome the limitations of our subjective intuitional capacities. It is true that the latter are little influenced by this process; but instead of constructing one single intuitive picture of the world, we learn to combine different pictures of different levels. Every picture may, besides containing false traits, introduce some true features into the composition. Perhaps it would be demanding too much if we insisted on including all features within one picture. The perspective of the beetle in the meadow is better than ours in the sense that it allows a more precise observation of the individual

[6] Cf. L. Fleck, *Entstehung und Entwicklung einer wissenschaftlichen Tatsache* (Basel, 1935), p. 147, Table III. Fleck shows antique and modern drawings of the human skeleton taken from medical textbooks; he makes clear that in ancient drawings the skeleton is always a symbol of death, whereas in the modern it is a symbol of mechanical-technical constructions.

blades; but the green evenness of the meadow which *we* see is an essential feature also, although unattainable for the beetle. When we see the polished table as a smooth surface, this is not simply false—this picture contains some qualities of the physical table which the picture of the swarm of gnats suppresses, namely, the relative smallness of the corpuscles and interstices compared with the two-dimensional extension. It is true that our substitute world is one-sided; but at least it shows us some essential features of the world. Scientific investigation adds many new features; we look through the microscope and the telescope, construct models of atoms and planetary systems, and penetrate by X-rays into the interior of living bodies. It is our task to organize all the different pictures obtained in this way into one superior whole. Though this whole is not, in itself, a picture in the sense of a direct perspective, it may be called intuitive in a more indirect sense. We wander through the world, from perspective to perspective, carrying our own subjective horizon with us; it is by a kind of intellectual integration of subjective views that we succeed in constructing a total view of the world, the consistent expansion of which entitles us to ever increasing claims of objectivity.

§ 26. Psychology

In the foregoing section we have shown how the external world is constructed on the concreta basis. It remains for us to show that the internal world also may be constructed on this basis. This means that we must show how so-called psychical experience is inferred from the basis of concrete objects.

In taking up this task, we depart from the traditional conceptions of psychology. It is the usual conception that so-called psychical phenomena are accessible to direct observation—that an internal sense shows us these phenomena in the same way that the external senses show us external phenomena. For a criticism of this conception we refer to our third chapter. We argued there that impressions are not observed but inferred; that we do not sense impressions but things; that there is no internal sense but that this concept is due to the confusion of an inference with an observation. We shall maintain these results now and apply them to a construction of the whole psychical world on the analogy of our construction of the external world.

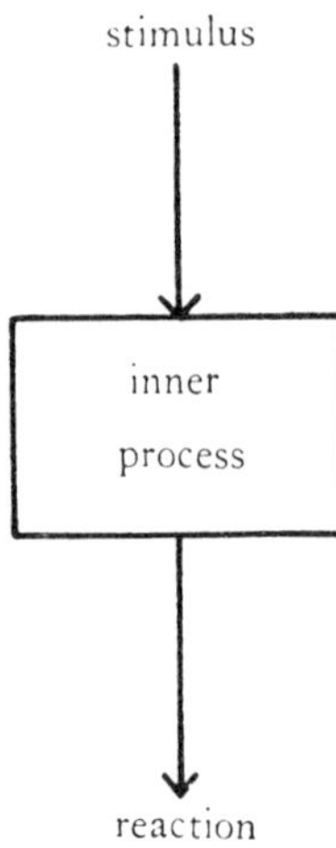

FIG. 5.—The human body as a system of inner processes intercalated between stimulus and reaction.

The human body is a system which is acted upon by external processes, and which itself initiates actions upon external processes. The external processes of the first kind are called *stimuli*, the external processes of the second kind are called *reactions*. Between them is intercalated the human body with its inner processes (cf. Fig. 5). The problem of psychology is to infer these inner processes. To illustrate this task, let us give an instructive example from physics which has been constructed, for this purpose, by Carnap[7]—an example which shows that the situation in question is not restricted to the case of the human body but occurs in a similar way for inanimate systems. A photoelectric cell is a device which is acted upon by light rays, as stimuli, and which produces an electric current, as reaction. In the interior of the cell there are processes; these, however, are not accessible to observation. In spite of this fact, a description of these inner processes may be

[7] *Erkenntnis*, III (1932–33), 127.

given in an indirect way. If there is a light ray of the intensity S falling on the cell, we may say that the cell is in the state corresponding to the stimulus S. Thus the cell is described by a description of the stimulus. A second way would be to describe the cell by a description of its reaction; if there is an electric current of the intensity R flowing from the cell, we may say that the cell is in the state corresponding to the reaction R. Both ways of description are equivalent, as there is a one-one correspondence between S and R.

What is important here is that we can give a very exact description of the internal state of the cell without being able to observe the interior. The best microscope directed to the interior of the cell would not show us any difference between two states S_1 and S_2, or R_1 and R_2; the inner changes are much too small to be observable. But the indirect description replaces the direct one to a high degree.

The situation of the psychologist is of the same kind as the situation of the physicist in the case of the photoelectric cell. He does not see the psychical phenomena but describes them by describing the stimuli which produce these processes, or the reactions which are produced by them. The idea of introspection is an illusion if we understand by introspection an observation of "psychical" phenomena; what we observe are physical phenomena, and the inner processes corresponding to them are only inferred. They are *illata;* and the basis from which we infer them is the totality of the concrete objects of the physical world, which stand to the inner processes in the relation of stimuli or reactions.

It is a current opinion among philosophers that what we have said is valid only for our observation of other persons, as we cannot share their psychical life, but that for our own person there is another means of observation, a direct view into our internal life. This distinction is one of the profound misunderstandings on which the traditional metaphysics is based. To clarify this question, let us enter into an analysis of the difference between our own personality and other personalities. There is, of course, a specific difference; but it is not of the type assumed by traditional philosophy.

We may begin this inquiry by a remark which a desultory survey of psychology already urges upon us: For the description of our own inner phenomena we generally start from the stimulus basis, whereas for the description of the inner phenomena of other persons we generally start from the reaction basis. I will illustrate this by an example.

The stereoscopic impression is a certain impression of spatiality which we may obtain from certain pairs of pictures if we observe them through a stereoscope. This impression demands, however, a certain training; untrained eyes have to make an effort before succeeding in obtaining the stereoscopic impression, and there are persons who never succeed in so doing. Looking through the stereoscope, we see two pictures at first; then these converge until they coalesce, and at this moment we see one, and only one, spatial picture in which the dimension of depth is seen in full strength, as in ordinary binocular vision of spatial things. The appearance of the spatial picture is rather sudden; the picture jumps suddenly into the spatial depth.

The description we have just given corresponds to what is called observation by introspection. If we analyze it, we discover however that it is built up entirely in terms of the stimulus sphere. A picture is a thing which we see, is a drawing on paper which we know to be an imitation of certain other physical things. The moving of the pictures is a physical phenomenon; so is the spatial depth—it is a quality observed in the visual perception of almost everything. There are some terms, in addition, taken from other phenomena of quite a different character, applied here in the sense of an analogy, such as the terms "coalesce," "full strength," "jump." Using these terms, we want to express a similarity relation between the objects just seen and other objects; "coalesce" indicates a similarity to certain changes occurring in the mixture of liquids, "strength" means a comparison to certain features observed in touching resisting forces, etc. We perform the description by describing physical things which stand in similarity relations to the thing observed—that is what philosophers call description by introspection. Our internal process "stereoscopic impression" is not observed directly; it is determined only as the internal process belonging to the stimulus S, where S is described in terms of concepts which a physicist would use for the description of a physical phenomenon.

Now let us see how we control the statement that another person has the stereoscopic impression. That the person is looking through the stereoscope is not a sufficient reason to believe that he has the impression. We control it by his reactions. First of all, we listen to what he relates. Speech is a special case of reaction but not the only one; and, above all, not always a reliable one. If the person says that he sees only one picture, and that it has a spatial character, this is not a sufficient indication that he really has the stereoscopic impression. It may happen that he neglects one of the pictures, i.e., drops it out of the field of concentration and sees then the other picture alone, mistaking the feeble spatial qualities of each photograph as the stereoscopic effect. In observing many persons before the stereoscope, I found a good means of eliminating this mistake. When the stereoscopic effect occurs, almost every person, especially if untrained, shows a sudden expression of joy and surprise, by an exclamation or a smile. This reaction, in combination with the other ones, is a very good indicator.

We see that here the presence of the stereoscopic impression is mainly inferred from observations of the reaction sphere. But not entirely; we observe also that the person has the stereoscope with the photographs before his eyes, i.e., we observe that a certain stimulus is acting upon him. But in this case—and this is the difference from the photoelectric cell—the occurrence of the physical stimulus S is not unambiguously combined with the inner process. This is the main difficulty of psychology. The same physical stimulus may start different impressions. We control the impression, therefore, by the reactions; only a rather complicated combination of stimuli and reactions allows a determinate inference as to the inner process. The decisive character here is always on the side of the reactions; they decide the choice between the different possibilities of impressions opened up by the stimulus. We may say, therefore, that the impression of another person is characterized by us as the internal process belonging to the reaction R.

The ambiguity between the physical object and the impression is the fact which led us, in the preceding sections, to the distinction of immediate and objective existence. The same objective thing may produce different immediate things. This was the case in our example of the juniper bush, which from a certain distance appeared as a man, from another distance as a bush. There are other examples in which the physical conditions do not change at all, whereas the immediate thing changes. A perspective drawing of a cube may be inverted, as psychologists show, so that the front and rear sides are exchanged. We know picture-puzzles in which suddenly the outlines of a man appear whom we did not observe before. All this means that the objective physical thing does not fully determine the immediate thing. As the impression is characterized by the immediate thing only, and not by the objective thing, psychology is interested in the description of the immediate thing.

This is the main difficulty of psychology. If the human body were organized like a photoelectric cell, psychology would be a very easy science; it would have nothing to do but to name the stimulus S, and would in this way describe the impression. The task of description in psychology, on the contrary, is to describe the immediate thing, not the objective thing.

This description, however, can be performed entirely within the stimulus sphere. We may refer here to the results of our preceding chapter; we describe the immediate thing by denoting objective things to which it stands in similarity relations. The different immediate things seen in a picture-puzzle are described if we say whether the picture resembles a man or not. If the indication of the stimulus S is not sufficient to determine the inner process, we may overcome this ambiguity by adding statements concerning relations of the type $S = S'$. In this way, all psychology can be presented within a stimulus language, i.e., a language using concepts characterizing things and relations of the physical world in so far as these occur as stimuli. Psychology, then, describes physical objects as well as physics does, but there is a difference in the aim of the description: physics looks for all those relations of a certain thing to other things which are needed for an unambiguous determination of the objective thing; psychology on the contrary looks for all those relations which are needed for an unambiguous determination of the immediate thing, and with this of the impression. It is another class of relations which psychology constructs; this is because psychology is not interested in the things but in the internal processes the things start in our bodies. In any case, these internal processes are not observed but inferred; the basis of this inference is observed concrete things.

We may also write a psychology in reaction language. For this purpose we should denote the internal process by giving the class of reactions belonging to one internal process. Linguistic utterances stand in the first place here; but they are not sufficient. As we saw in the example of the stereoscopic impression, additional reactions of other types are required, such as exclamations, movements of eyes and face, etc. Reaction language is used in the psychological observation of other people and of animals; we shall inquire shortly for the reason.

The relating of both systems of description is one of the most important tasks of the psychologist. Which description of the reaction sphere belongs to a determinate description of the stimulus sphere? This is one of the main questions of psychological investigation. The relation of S to R is, for the human body, a very complicated matter; it can be answered only if we consider not isolated stimuli and reactions but rather comprehensive groups of them. We put a piece of sugar before a dog; will he eat it? Perhaps this reaction depends not only on the food stimulus but also on other stimuli given in the form of gestures by the dog's master. We tell a man that he is to be put into prison; will he run away? That depends on many other conditions, such as the crime he is charged with, his knowledge as to the conditions of life in prison, his chances of further life after escape. It is always the relation of S to R which is asked for in such questions.

Now, if we are to compare both languages, we have to note a decisive difference. A complete description of the inner process in stimulus language without the use of reaction language can be given only by the person himself whose process is to be described. A complete description in reaction language, however, is obtainable for every other person. This difference is caused by the variability in the relation of the stimulus to the impression. The immediate thing is observable only by the man whose impression is asked for; only he can say whether he sees the juniper bush or the little man with the caftan, or whether he sees a man or not in the picture-puzzle. Other people depend upon his reactions, which may consist in his linguistic report or other indications.

This distinction is due to the special position of the self-observer. In the case of self-observation the man who observes is identical with the system the inner process of which is to be characterized. Now, *seeing* the immediate thing is identical with *having* the corresponding inner process; therefore the man who has the inner process observes the immediate thing—nobody else is in an analogous condition.

It is to be remembered here that *having* the inner process does not mean *observing* the inner process but means observing the immediate thing. This matter is the source of much confusion and, I think, of the false conception of the "psychical phenomena" of traditional philosophy. We do not see our interior process, but we have it; and, because we have it, we see a thing outside. By a confusion of these relations the idea has arisen that having an inner process means observing it, and "having" has acquired the sense of "being given in observation." But "having" is to denote here only that the internal process occurs within our bodies. If this is the case, we observe a thing outside. However, this is an immediate thing; without further determinations we cannot say whether it is jointly an objective thing, or whether there is only a coupled objective thing, or whether there is no co-ordinated objective thing at all, such as in the case of dreams.

The special position of the self-observer has led to the concept of introspection. If this term is to denote nothing but that the self-observer is the only one who can give a complete description of the immediate thing, without making use of reactions, the term would be permissible. The term has been connected, however, with the idea of a direct observation of inner processes; so it has acquired a misleading metaphysical meaning. We shall therefore avoid this term and replace it by the term *self-observation*.

The idea of introspection has been developed, I think, in a misinterpretation of a fact which indeed offers the possibility of misunderstanding: it is the fact that the stimulus may be situated within our own body. We have already discussed this case in our criticism of impressions (§ 19); we held there that, as well as seeing our body, we may feel it by interior tactile sensations, and we added that this sensational character is revealed in the fact that such inner sensations are always spatially localized. We must extend this criticism now to the more general case of the so-called psychical phenomena of a higher level, such as thoughts, emotions, passions, etc.

It is one of the arguments in favor of "psychical experience" that these phenomena have no localization; Kant already took as a specific quality of psychical life the supposed fact that it passes within time only but is not localized in space. I feel, say, a certain joy at a definite time; but this joy has no place in space. I had the thought of going to the cinema last night at seven o'clock; but this thought had no position in space. Psychical phenomena such as love or hatred may last for a period of time, for hours or years; but they have no spatial extent. This nonspatiality of psychical life is considered as one of the most cogent arguments for the duplicity of our experience which divides, it is said, into the two domains of physical and psychical experience; the former is ordered in space and time, the latter in time only.

This theory, it seems to me, is the result of a twofold confusion. First there is the confusion indicated concerning stimuli situated within our bodies; these stimuli are considered as entities of a nonphysical character. To this confusion is added a second which arises from the problem of abstracta.

Joy, grief, love, hatred, etc., are abstracta, complexes of elementary phenomena which are "bodily feelings." Our bodies feel light, without weight; we feel ourselves "walking on air," smiling—of such a kind are the elements of the complex called "joy." We feel a certain tension within our bodies, a constraint to move and see a certain person, we feel our body becoming more vivacious in the presence of this person, feel excitations in the sexual zones of our body—of such a kind are the elements of the complex love. These elements have spatial localization, either in special parts of our body or all over the body. The abstractum, the complex, may however be defined in such a way that it has no spatial qualities at all. We discussed this question in § 11 and gave examples of abstracts composed of physical elements but having no spatial qualities, such as the political state, a melody, or the elasticity of a spring. We said that this matter depends on a convention, that we may or may not ascribe a spatial position or extent to these abstracta, but that usually the nonspatial conception is preferred. This is valid as well for the complexes composed of stimuli situated within our body. The complex "love" is generally conceived as having no place or extent in space; but we might give another definition according to which this complex is spatially situated within our body and extended all over our body. The preference of time in these cases, the decision for a localization of the abstractum in time but not in space, has its origin in the fact that the temporal characterization enables us to construct an order among the abstracta, by ascribing different time positions to each of them; whereas a spatial characterization leads to the trivial result that they are all within our bodies and extended all over it—thus making it impossible to establish an order among them. If the same result occurs for a position in time, we arrive at a similar indetermination for the time qualities of the abstracta. Is the character of a person *in time?* If this is assumed, the character covers the entire span of the person's life, and therefore this definition is of no use for establishing an order. We cannot establish a time order between his character and, say, his father-complex, an abstractum which also covers all his lifetime. For these kinds of "psychical phenomena" the temporal characterization is usually dropped—an indication that temporal character is not at all serviceable for the definition of "psychical experience."

The fact that the stimuli may be situated within our body has a consequence of quite another type than is assumed by traditional psychology. What it implies is that in these cases the inner processes are for us concreta. We said previously that inner processes are illata; this is valid for the processes of optical and acoustical sensations, which are inferred from the observation of external things. The inner process "hunger," on the contrary, is a concretum; it is directly observed in the same sense that we observe, say, a movement of our legs with the tactile sense, or the pulsation of our heart. The interior of the body is partially accessible to direct observation, partially only inferred—as is the case with most external objects. The abstracta composed of complexes of these internal concreta and illata constitute the so-called higher psychical life.

The internal processes are, to summarize, inferred from stimuli or reactions, or observed by the inner tactile sense. What then are these internal processes, if we are to ascribe to them a place in our physical world?

They are nothing but physiological processes. There is a direct way to observe all internal processes of the human body; this is the way of the physiologist. He discovers that an optical sensation consists in a picture on the retina, in determinate physiological changes in the nervus opticus and the brain; he finds that hunger consists in convulsions of the stomach, secretions of the salivary gland, etc. He is not bound to the stimulus language or to the reaction language; he observes the interior of the bodily system directly and expresses his results in a direct language, which we may call *inner-process language*.

There is an old question which has been opposed at all times to materialism: How is a nervous process in the brain transformed into an optical sensation? How is a convulsion of the stomach transformed into the feeling of hunger? This question, I think, is nothing but a profound misunderstanding of scientific concepts. Let us analyze the questions separately; they are of different types.

An optical sensation is not observed by a man who sees things outside his body; it is inferred. The man *sees* a thing before him and *has* a sensation; this sensation is for him an illatum. He does not know anything about its qualities, except that it has a certain correspondence to the immediate thing he observes. It is an unknown, X, determined as a function of the immediate thing observed. If now a physiologist asserts that this X is a nervous process, there is no difficulty in characterizing X as a process in the nervous system. There is no more difficulty in this than in a similar case of the physical world, say in our photoelectric cell. The internal state of the cell has been first determined as the state X belonging to the intensity S of the light ray entering the cell; later the physicist discovers that the state X consists in a certain swarm of electrons passing through the spaces between the molecules of the photoelectric crystal. The physicist does not ask: Where is the light ray within the crystal? How is the swarm of electrons transformed into a picture of the light ray? These would be unreasonable questions issuing from a misunderstanding of the functional relation between the light ray, as stimulus, and the swarm of electrons, as the internal process released. The light rays coming from the external thing release the nervous process within us. It would be unreasonable to demand that this nervous process is to be transformed into a picture of the light ray, or of the external thing. Having the nervous process means seeing the external thing; from this we cannot infer that the nervous process is a picture of the external thing, or is transformed into such a picture.

In our second example, the sensation of hunger, the situation is a little different. In this case the internal process is in itself observed by us. We do not sense it as a movement of our stomach, as the physiologist describes it. But this is a difference we notice similarly in cases of external observation. We see a rectangular box as a geometrical body with planes, edges, and points. If we touch it, we feel it as a resistance, we feel the sliding, cutting effect of the edges and the stinging pressure of the corners on our fingers. This difference of qualities is due to the difference of the sense organs used in the observation. Correspondingly, hunger observed by the inner tactile sense has qualities different from hunger observed with the eyes, as a convulsion of the stomach. Similar differences occur within optical sensations in the form of differences of perspective; the view I have of a certain room differs from the view another person has. In this case, an exchange of spatial positions is easily performed, and the other person may have my perspective also. In the case of the observation of internal processes of the body, however, an exchange of positions is physically impossible. If a physician who watches my hunger on the Roentgen screen should want to feel the hunger I do, he would be obliged to enter into the same tactile relations to my stomach as I have—this is physically impossible.

The difficulties of the problem of internal processes arise from the fact that there are three different ways of determining these processes: the way of observing the stimulus, that of observing the reaction, and that of direct observation of the interior of the body. The latter divides into the two ways of physiological observation and of self-observation by the inner tactile sense; the first of these is open to every person, the second only to the person who is identical with the body in question. The difference in the ways of determination has led to the idea of different objects concerned. This is the decisive fault; all methods in fact have the same objects.

Traditional psychology prefers throughout the stimulus method and is accordingly written in stimulus language. To this is added the method of self-observation of the body by the inner tactile sense; but the main role is played by the stimulus method. This is because most of the "higher psychical phenomena" are produced by external stimuli and therefore best described in the stimulus language. The immediate thing is described by comparisons to other physical things of a similar kind. We speak of the "stabbing pain" we had on hearing the message of the death of an intimate friend and describe the immediate thing "pain," furnished by the inner tactile sense, by a similarity relation to the immediate thing "needle" which we may feel stabbed into our finger. We say, "I felt bound to go to my friend," and compare the tension felt in our muscles with the sensation of a cord bound round our arms. We talk of a man who has "a clear insight into his task" and describe the subjective images he has of his future work by a comparison to optical qualities of bodies seen in bright light and a clear atmosphere. This method of description by comparison in the stimulus language is also the method of poets.

> My heart aches, and a weary numbness pains my sense
> As though of hemlock I had drunk
> Or swallowed some dull opiate to the drains
> One moment since, and Lethewards had sunk.

These verses of a romanticist—quoted from Keats's "Ode to a Nightingale"—give a description of a psychological state in the stimulus language. The feeling is described as that impression which occurs after drinking hemlock, or an opiate; the "aching heart" is a description of a feeling such as appears after our body has been injured from without, or such as is observable as released by internal stimuli. Only the term "Lethewards had sunk" belongs to reaction language, as it describes a reaction occurring in combination with feelings of the type indicated. Reaction language is generally used in poetry if the poet wants to describe a person in an objective way, i.e., if he wants to prevent us from identifying ourselves with the person. "You are fatal then when your eyes roll so," says Desdemona; the poet here wants us to see Othello through the eyes of his wife.

The behaviorist, in opposition to the traditional psychologist, considers the reaction language as the only language of psychology. That is to say, a behavioristic description includes the stimulus, but only in its objective physical existence, not in its immediate existence. As the inner state of the person is not determined by the objective stimulus, the determination of the inner state is entirely left to reactions; thus reactions are considered as the only indications permissible in psychology, and in this sense the language of behaviorism is reaction language. The relation from *S* to *R* is what the behaviorist studies; *S* characterizes the environment, *R* the person or animal with all his inner qualities. To this is added, in a certain degree, the inner-process language in the objective, i.e., physiological, form. The limits between the objective inner-process language and the reaction language are fluctuating; it is not always sharply demarcated where the inner process ceases and the external reaction begins. Some processes within the body are usually called reactions, such as palpitations, blushing, etc.; they might be considered as parts of the inner process as well. The behaviorist usually considers only those inner reactions or processes which are easily observable from without, such as those already mentioned; processes demanding, for observation, operative intervention, e.g., processes within the nervous system, are left to the physiologist. Here also the limits are indeterminate.[8]

It is the advantage of behaviorism that an objective language is obtained which can be controlled by everybody; reports of the person observed are not needed, and the method is applicable to animals as well as men. Restriction to this method, however, seems to be an overstrained requirement. This postulate arose from an antagonism to vague metaphysical concepts in traditional psychology and had, therefore, a methodological value in the sense of a strict purification of psychology. It seems to me, however,

[8] The Pavlov experiment on the salivary gland demands, for animals, a simple operation but is used by behaviorists also.

that to lay aside the reports of the person observed is to eliminate the most privileged observer. We know that subjective reports are sometimes dubitable, and the elaboration of methods of control is very useful. But the unique position of the self-observer offers such great advantages that psychology will never, I think, renounce using it. It is the fact that the self-observer, and he alone, can describe his internal state in stimulus language, without the use of reactions, which makes this position unique. A man who sees a juniper bush, at nightfall, as a brigand, knows this and does not need to infer it from his palpitations or trembling knees. A man who has hunger knows this from direct sensation and does not need to count drops from his salivary gland. There are a great many psychological facts which never would have been discovered without the self-observer.

Take as an example the fact that we see parallels, such as rails, converge. It is a subjective fact, since the objective physical stimulus does not give any indication as to this psychological fact. It is, however, easily described in stimulus language: "I see these rails similar to such lines," and with this the person points to a drawing of convergent lines. I do not see any way in which this psychological fact might have been discovered without a report of a self-observer. I do not say that it is absolutely impossible to discover such a fact by behavioristic methods but only that this is out of the domain of the practically attainable. The report of the self-observer is in a great many cases a means far superior to the observation of reactions.

It is true that the report, as soon as it is uttered, is in itself a reaction. But the question is precisely whether the behaviorist is to include report reactions. That the knowledge of the person observed, if it is to be transmitted to another person, must be transformed into a reaction, is obvious. But if the person observed wants to know what he himself observes, he need not wait for his own reaction. He may even suppress his reactions and keep his knowledge to himself. The cardplayer knows what he is hiding behind the poker face. If psychologists had none but persons of the poker-face type as subjects, they would have a very difficult task.

Behaviorists may answer that thinking is subvocal speaking, that a man who knows what he observes speaks to himself subvocally and therefore also knows it from his reaction as other persons do. This objection, however, would not correspond to thoroughgoing behaviorism and would not, I think, be shared by Watson. For behaviorism, subvocal speaking *is* knowing; so the man does not obtain his knowledge from subvocal speaking. He obtains it from seeing objects, i.e., in physiological language: the nervous process of seeing releases subvocal speaking. Other persons, however, remain one step behind: their knowledge, i.e., their subvocal speaking, is started by the vocal speech of the self-observer.

The method of self-observation is, I think, a necessary element of psychology; it is to be controlled but not to be dropped. The mischief of psychology does not arise from this method but from the false interpretation which has been given to it. It is the concept of introspection which marks this misinterpretation, as it is meant to indicate a direct view of psychical phenomena. The interpretation developed by us, in the sense of a stimulus language, is free from such misconception. The case of the converging parallels gives a good example of a psychological description in stimulus language. What is stated here is a comparison of two objects: the rails, which are physically parallel, and the lines drawn on the paper, which are physically convergent. By this comparison, the immediate thing "rails" is described, and with this, indirectly, the inner process "impression." By this method we can describe our impression even to a man born blind. The method of self-observation, if it is conceived as the method of stimulus language, is not less objective than reaction language. However, it opens up possibilities for observation which do not exist for the reaction method.

Our solution of the problem of psychology is based on the distinction of the three categories of stimulus, inner process, and reaction; to this is to be added the fact that the self-observer is in a particular position which cannot be occupied by other persons. We must now add a remark concerning the relations between the three categories.

These relations are generally considered as implications; the stimulus implies the inner process, and the inner process implies the reaction. It is the same case as in other causal relations; the light ray implies the inner state of the photoelectric cell, and the inner state of the cell implies the current leaving the cell. But, just as in all these other cases, this is to be considered as an idealization; the relations are, strictly speaking, not logical implications but probability implications. That is to say, if there is a certain stimulus, then there is a determinate probability that a certain inner state will occur, and, if there is a certain inner state, there is a determinate probability that a certain reaction will occur. Even in the case of the photoelectric cell there are, strictly speaking, only probability implications; in the case of the human body this is more important because the degree of probability obtainable is not so high as in the case of the cell. The intervention of the probability concept in this context adds some relevant features to the problem of psychology.

The first consequence is that the inner state of the body cannot be conceived as a reducible complex of the stimuli or of the reactions. It is, on the contrary, a projective complex of these elements. This distinction introduces into the problem of the psychology of other people a remarkable correction.

Behaviorists used to say that what we mean by speaking of the psychical state of other persons is just the class of their reactions. If we say that a man is angry, this means —so behaviorists argue—that the man speaks in a loud voice, springs from his chair, and, leaving the room, slams the door. This conception, however, is not tenable. A statement about the reactions as described is not equivalent to the statement about the anger but is in a probability connection only. This is important as to the bearing of behavioristic methods. Psychologists frequently show a deep-rooted aversion to behaviorism; they will not admit that speaking of a man's fury means speaking about his visible reactions, but maintain that what they mean is something else which they infer only from the reactions. This objection, I think, is right. It is confirmed by our probability theory of meaning.

What then is the meaning of our statement about anger? This is asking for those elements of which fury is composed as a reducible complex. The answer is that these elements are given by the internal physiological state.

Indeed, if we know all the visible reactions of a man, we may infer with probability only that he is in the internal state called anger; but if we knew his inner state exactly, including all processes in the nervous system, the question of whether he is angry would be decided. The definitions of psychological states are to be given in the form of descriptions of inner processes. If we replace them by descriptions of certain stimuli or reactions, this is to be conceived as a practical abbreviation which is valid only in the sense of an approximation.

This is the reason that psychology so frequently stands before questions unanswerable in practice. The probabilities of the implications from behavior to inner states are, in many cases, not very high; thus the psychologist cannot overcome a certain indeterminacy in all his laws. I do not mean to say that all progress is precluded; but a determinacy corresponding to physics will be reached only if the direct physiological consideration of inner processes is achieved in a much higher degree than it is today. This remark, however, is valid in principle only. In the present state, on the contrary, as physiology is not yet able to distinguish internal states in such a degree of precision as is furnished by the observation of stimuli and reactions, the description of the inner states by means of the stimulus and reaction language is much more exact than the physiological description. This is the reason that psychologists refuse physiological methods and keep to self-observation and observation of reactions. The psychoanalysis of Freud, for instance, which is formulated entirely in stimulus and reaction language and does not use physiological language at all, gives very deep insight into certain internal states, such as "complexes"; physiology is by no means able to give the corresponding physiological descriptions. This is why psychoanalysis is used as a special medical method in cases in which those of physiology fail.

If to our distinction of the three categories of stimulus, inner process, and reaction we now add the fact of the probability character of the relations between these categories, the task and method of psychology assume a rather complicated character, but one in its general structure of a type similar to that of physics. Psychology is a science which infers illata from concrete objects. The inferred objects are projective complexes of these concrete objects. Since some of the objects of psychology such as bodily feelings are accessible to the inner tactile sense, the inferred illata in such cases are internal elements of the observed concrete objects; it is therefore the process of internal projection which plays a role here. The "higher" psychological objects, and just those most frequently occurring in practical psychology, i.e., psychology as needed for daily life, are abstracta, built up of concreta and illata.

This characterization of psychology needs no such thing as "psychical experience" and is therefore very different from the usual metaphysical conception of psychology. On the other hand, behaviorism appears as an oversimplified conception, which, it is true, avoids metaphysical misinterpretations, but which does not take into account two remarkable facts: the particular position of the self-observer and the probability character of the relations between the three categories.

If we compare the process of the construction of the internal world to that of the external world, there is no difference in principle. The basis in both cases is constituted by concrete objects, including in this class objects both outside and inside our body. The construction of the external world is performed by the addition of objects outside our bodies, obtained by projections. The construction of the internal world is performed by the addition of objects inside our bodies, obtained, for the greater part, by projections. The first case is conformable to common sense; the second may appear strange and circumstantial. This may be the reason why the idea of a direct view into an internal life was invented. This idea, however, is not tenable. Our knowledge of the internal world is obtained by inferences which are based to a great extent on phenomena outside our body. It is as though a motorist were to infer a rising temperature of his motor from the steepness of the road his car is mounting.

§ 27. The so-called incomparability of the psychical experiences of different persons

Let us apply our results concerning psychology to a problem arising within this domain and frequently discussed in philosophy.

There is something in our experience, so it is said, which is accessible only to ourselves, and which cannot be communicated to other persons. We see the color red, we feel the heat, we taste the sweet; but we cannot tell how we see or feel or taste it. Other people tell us that they also see the red and feel the heat and taste the sweet; but we never can compare these sensations with ours, and so we do not know whether they are the same. There is, therefore, an unutterable residue in our experience. This is one of the most frequently used arguments in favor of the existence of a particular psychical world within every person; this world is supposed to be known only to each person and not accessible to others.

Let us analyze this situation. It is in a certain sense true that impressions of different persons cannot be directly compared. Imagine a man who sees green when I see red, and red when I see green—would we ever know this? A mind untrained in philosophy might perhaps object that the man in question would be in permanent conflict with the traffic regulations when driving a motorcar, that he would cross the street at the red light and stop on the green light—but of course this is thoroughly false. This man has learned that the color he sees when the red light is on means to stop, that this color is called "red," etc.; so all his reactions will entirely correspond to those of a man of "normal" impressions. There is no possibility of detecting the abnormality of this man.

This fact, however, is just an indication that the comparison intended constitutes a pseudo-problem. Neither for physical truth meaning, nor for probability meaning, nor for logical meaning can the comparison of the impressions of two persons be accepted as a meaningful question. This is not surprising, since even for the same person there is an analogous pseudo-problem; as we pointed out previously (§ 21), nobody can directly compare his impression of today with his impression of yesterday. The idea may still be generalized, and the case of psychological comparisons may be considered as a special case of a general physical theorem. We cannot compare the length of a meter bar, situated at one point, to the length of another meter bar, situated at another point; we cannot compare the second indicated by a watch to the following second indicated by the same watch. We need not enter here into a criticism of this problem, as it has been solved within the philosophy of space and time.[9] The indetermination in question, as it is shown there, leads to the consequence that in such cases it is not a *cognition* which is to be demanded but a *definition*. The equal length of two meter bars at different points of space can only be defined; i.e., if these fulfil certain observable conditions of another kind, such as being equal when they are put side by side at the same place, being of the same temperature, etc., we call them equal when they are situated at different places. In the same sense, the comparison of the impressions of two persons is a matter of definition. Here also the definition will demand that some observable conditions be fulfilled if the equality is to be postulated. If all reactions of the two persons, including reports of self-observation in stimulus language, are the same, we may define their impressions as being the same. It is only when such a definition has been given that the question of the sameness has a meaning; without this definition, there is nothing asked at all when

[9] Cf. the author's *Philosophie der Raum-Zeit-Lehre* (Berlin, 1928), §§ 3–8.

we say, "Are the impressions the same?" We must first co-ordinate with the term "same" a corresponding set of observable relations; only thus does the question become determinate. Definitions of this kind have been called, therefore, definitions of co-ordination.[10]

If such a definition is once given, the question of the sameness of impressions can be answered empirically. We may say that a color-blind man does not have the same impression of certain colors that other persons have but that normal persons have the same impressions. This "sameness," however, has only the meaning established by the definition, not an absolute sense.

It has been argued that an absolute comparison of impressions is not logically impossible, that it is only because of the limitation of our technical faculties that we cannot make such a comparison. Biologists[11] have succeeded in joining salamanders by an operation in such a way that they have a common circulation of the blood and even a common nervous system; the possibility cannot be excluded that some day the same operation will be successfully performed upon men. In such a case, one person could look through the eyes of another person. Let us analyze this idea.

Imagine two men combined in such a way that the nervous processes of one enter into the nervous system of the other. They stand back to back; before A there is a red light which A sees and calls red. B sees the light also, but by the eyes of A, as his eyes are not turned toward the light; B says, however, that the light is green. Now both persons turn, and the light stands in front of B; B now calls the light red, whereas A now calls it green. Would not this indicate an absolute difference of their impressions?

It would indicate a difference but not an absolute one.

[10] Zuordnungsdefinitionen (cf. *ibid.*, § 4).

[11] Cf. I. Schaxel, "Das biologische Individuum," *Erkenntnis*, I (1930), 467.

The statements made by A and B here presuppose already a definition of comparison. It is not true that the impressions of A and B are directly compared. Each one compares his present impression with a present recollection image of a previously seen object. When, for example, in the second position, the person A says that his impression is different from the impression in the first position, he compares not these impressions directly but only the recollection image of the first impression to the second impression. But then does he know which of the two has changed? What if the recollection image has changed and is different from the first impression, whereas the direct impression is unchanged? Then the impressions of the two persons would not differ. We see that such a comparison has a meaning only after a preceding definition and is therefore relative in the same sense as before.

We may, however, include the case of the combined nervous systems in our definition and say: Two persons have the same impressions if, first, they always show the same reactions and, second, if in the case of combined nervous systems, it makes no difference to them whether they look through the eyes of the one or of the other. The addition means that the experiment as described should furnish the opposite result, that if A calls a color "red," B calls it "red" also. If we use this definition, the question whether different persons have the same impressions cannot be answered with certainty but is a meaningful problem. It can, however, be answered with probability; we may say, I think, that it is highly probable that normal persons have the same impressions. This means it is highly probable that if two persons always show the same reactions, they would, after a combination of their nervous systems, discover no difference if they look through the eyes of the one or of the other.

We see from this that the sameness of the impressions in the narrower sense of the second definition has not only logical meaning but also physical probability meaning. It may be, therefore, admitted for our world. This definition seems to underly the ideas of such philosophers as want to maintain that a comparison of impressions means more than a comparison of reactions. Such an idea, we see, can be admitted, even for our world, if we accept probability meaning. But it is, of course, no absolute comparison; it presupposes also a definition of co-ordination, as all physical comparisons of this type do.

After these considerations, the problem of the incomparability of the impressions of different persons assumes an aspect very different from the usual view of the problem. This incomparability is not due to the individual separation of different persons but to a logical indeterminateness of a more general character, occurring in the same way for comparisons of purely physical character: this is the indeterminateness of the comparison between things or states in different spatiotemporal points—as is well known in the philosophy of space and time. This highly general character of the problem has been disregarded, and the incomparability of impressions has been considered a proof for the monadic character of the human mind. However, if we call the impressions of two persons incomparable, we are obliged to call the impressions of one person at different times incomparable as well. The analysis of the general problem, in the theory of space and time, has shown the means for surmounting these difficulties: a comparison can be made if we overcome the indeterminateness by the introduction of definitions of co-ordination. This principle is applicable for the comparison of impressions as well. If we introduce such definitions, the comparison of the impressions of one person at different times becomes meaningful;

but then the comparison of the impressions of different persons becomes meaningful as well and cannot be called impossible. The isolation of the human monads is, logically speaking, not of another type than the isolation of the different events within the stream of experience of one person. The difference is that, within one person, the phenomenon of recollection images furnishes a simple mechanism upon which a definition of comparison can be based, whereas for two persons, if all our requirements for such a definition are to be satisfied, a crossing-over of the nervous systems ought to be accomplished. Such an operation is as yet not technically possible; but it is not logically excluded. Its result, however, can be foreseen with some probability. Thus probability opens a window between the monads even if there is no channel uniting their individual streams of experience.

There is an outcome of the usual erroneous conception of the problem of incomparability which we must now discuss: it is the idea that there is something inexpressible in our experience, known to us alone but not communicable to other persons. The structural relations between impressions have been distinguished from the specific *quale* of each of them; only the structural relations, it is said, are communicable; the quale is known only to ourselves. The fault of this conception, it seems to me, lies in the idea that we ourselves know more than structural relations. We see differences between red and green; but to say that we see, in addition, a specific quale of the red means nothing. Such a term is nothing but a misleading expression for the fact that we can recognize red colors, i.e., that we observe them as the same. The relation of sameness has been substantialized—turned into a certain substantial entity called the quale, a fallacy frequently occurring in logic. If we had no possibilities of observing similarities, i.e., if there were no two similar impressions in the whole stream of experience, the idea of a specific quale would not have arisen. To realize this we must remember that, in this case, recollection images would be excluded; the capacity of memory to "preserve the quale" is nothing but the capacity for producing images which stand in the relation of sameness to observed things. That the quale is not permissible is shown also by another reflection. We talked previously of a man who has the quale of red and green exchanged, i.e., who sees red when we see green, and vice versa; we said that this exchange cannot be discovered, as the structural relations are the same for him and for us. Now imagine that the same exchange happens for us, that one day we see as usual, the next day with exchanged colors, the following day as the first day, etc. If this exchange affects our recollection images as well, we never should become aware of it. We should believe then in a constant quale of our impressions, whereas this quale in fact always changes. This shows that the quale is an untenable concept. Its tenable basis is nothing but the relation of sameness, and the term "quale" means as much as can be said about similarities.[12]

For an illustration we may refer once more to an example chosen from the theory of space and time. The idea of the quale may be compared to the idea of an absolute size in space, and is therefore exposed to the same criticism as this untenable concept. Our argument concerning an unobservable change of the quale from day to day would correspond to the well-known argument that nobody would be aware of the change of "absolute size" if, during one night, all things (including our own bodies) would be enlarged to ten times their size; just as these reflections

[12] It is no objection against our reasoning that we make use of the concept "quale" which we want to refute. Our method is the *reductio ad absurdum:* we presuppose there is a specific quale and show then that this presupposition leads to contradictions.

demonstrate that all we mean about spatial size reduces to relations between spatial things, the corresponding reflections as to an unobservable change of the quale demonstrate that it is only relations between observed things which we can "mean" and not an "absolute quale." Even for ourselves, the occurrence of a certain quale would not be verifiable.

What we know can be said, and what cannot be said cannot be known. The idea that we know more than we can say has its psychological origin, I think, in a certain psychological fact concerning the capacity of imagination. We can imagine things we have not previously observed, but there are certain limits set to this power. As to geometrical arrangements, there is, it seems, no limit for imagination; but there is a limit as to colors, tastes, and some other qualities. We can imagine an elephant with six legs, though we never saw one; but we cannot imagine a color outside the well-known domain of usual colors. This is the reason we cannot describe to a color-blind man the colors we see. Suppose we show him a set of differently colored objects, but all of the same intensity. He will see them all equally gray, whereas we see differences among them. We can say to him: this thing is, for us, equal to this, but this thing is different from both. He may believe us, but he cannot imagine that there is a difference. If he could, he might attach the imagined difference to the things; he would represent then, for himself, differences which he did not see. It would correspond to the case when we look at two elephants and imagine that one has six legs; though we do not see such a difference of the elephants, we could imagine it. Now suppose the same power of imagination for the color-blind man; though he sees no differences of colors, he might imagine them and in this way construct a colored world of his own. Would this be the same as our colored world? This, we found, is an unreasonable question; if his world has the same structural differences as our own, it may be called equal to our world. We are right, therefore, in saying that in such a case we had described the colored world to a color-blind man—though he would continue to be unable to see, in given physical objects, the color differences we see and would not be able to drive a car according to the directions of the traffic lights. Only imagined things would show color differences for him; but, as to observed physical things, he would not know where to attach the differences he could imagine.

This expansion of the observed colors by imagination is, however, impossible. It is this limitation of the power of imagination which leads to the idea that there is something inexpressible in our experience. We say: Whoever wants to know what is red must look at a red thing. But we do not say: Whoever wants to know what is an elephant with six legs must look at such a thing. The red, therefore, is called an *inexpressible quale;* the six-leggedness is not. This is a rather incorrect mode of speech. We ought to say: There are certain differences which we cannot imagine without having seen them before. It is a certain indigence of fancy which we have to state here—no more. It is true that we cannot describe colors to a color-blind man; but this does not mean that what we know about colors is unutterable—it means only that the color-blind man cannot imagine certain differences which we see and which we describe to him. The existence of limits of imagination[13] in certain domains, together with a false theory of the comparison of impressions, is the origin of the untenable idea of the inexpressible quale.

[13] It would be an interesting task for psychologists to find out whether these limits are so rigid as is usually assumed. It may be that with training an expansion of color imagination is attainable, for color-blind people as well as for other people.

A third source of this conception may be indicated. Suppose a color-blind man who possesses—in opposition to usual experience—the capacity of imagining color differences in the example just cited. Suppose besides that some day physicians find an operation which gives to our color-blind man the capacities of normal vision. Will the colors he then sees correspond to those he had imagined?

This of course cannot be guaranteed; it may be that the new colors are entirely different from the imagined ones. Philosophers may accordingly argue that this proves the existence of the quale: we could not describe this quale to the man, and he had to learn it by his own experience, made possible in our supposed case by an operation.

We cannot however accept such an argument. What is to be said here can be said entirely by means of similarity relations. The new colors are not similar to the imagined ones—this is what the man observes. Such an experience, however, may always happen. We have no guaranty that the colors we shall see tomorrow will be the same as those seen today. It is the indeterminacy of future observations which enters here and which furnishes a new source for the idea of the inexpressible quale. But it is to be realized that nothing more is involved than the occurrence or nonoccurrence of similarity relations.

A word may be added. Similarity relations permit predictions; thus we may say: If you look at this body tomorrow, you will see a similar color. In the case of our color-blind man, we cannot make such a prediction; i.e., we cannot say: The color you will see after the operation will be similar to the color imagined before it. The difference is nevertheless only a difference in the weight of a prediction. The second prediction is meaningful but is likely to be false. There is a natural law which we previously called the constancy of the perceptual function; it enables us to make accurate predictions, by means of the similarity relation, of future observations in comparison to past ones. There is no such law as to the comparison of imagined things and future observations. If the imagined thing can at least be put into some relation to formerly observed things of a different kind, there is a certain approximation possible. We can describe to a person the color of a flower he never saw by comparison with colors of a somewhat different kind; we say, for example, "A deeper violet than this, and tending more toward red." In this way, we may obtain a rather reliable prediction. In the case of our color-blind man, we cannot predict a similarity relation between his imagined colors and his future color observations because we cannot show him, before the operation, physical things which, for him, will be similar to his future observations after the operation. This expresses, however, nothing but a lack of determinacy between his observations in so far as they are separated by the operation.

What stands in the background here is the fact that an observation is always imposed upon us, that we do not produce it but receive it independently of our own wills. We shall speak of this passivity in observation later on (cf. §§ 30 and 31). It may suffice to say here that this idea is sometimes expressed by saying that observation furnishes the quale of the impression. Nevertheless, this is a rather misleading term. Observation furnishes the whole impression, and whether it is similar to a former one, and in what respect, cannot be foreseen with certainty. This is all that is involved; we need no such quale as metaphysicians have invented.

§ 28. What is the ego?

The question as to the difference of the impressions of various people leads us to another question concerning the special position of ourselves in the world; this is the question, What is the ego?

Metaphysicians of all times have written much about the ego. They have insisted that it is the cardinal point to which to attach all knowledge about the world, that the ego is a metaphysical entity known directly to ourselves, that it is a "thing in itself" but known to us by way of exception—and many other doctrines which under the scalpel of exact analysis turn out to be nothing but metaphors camouflaging a lack of insight into the logical nature of psychological phenomena. Our analysis of psychology furnishes an answer of quite a different type: The ego is an abstractum, composed of concreta and illata, constructed to express a specific set of empirical phenomena.

Let us collect these phenomena. Our characterization of the specific position of the self-observer furnishes the way to point them out. First is the fact that among all human bodies there is one, our own body, which accompanies all phenomena. We see the table and the paper on which we write, and there is one hand, our hand, on this table. We can turn our heads in such a way that the hand is not seen—then we still feel the existence of this hand by the tactile sense. We cannot rid ourselves of this world of bodily feelings. We observe that they are connected with certain other phenomena; when we see a needle stabbed into our hand, we feel it, whereas we feel nothing when we see the same needle stabbed into the hand of another person. We desire to move our legs, and we do so immediately; but we cannot move the legs of other bodies in such an immediate way. Thus our own physical body appears to be in a unique relation to a set of observed phenomena.

There is, second, the fact that some physical phenomena are known to ourselves alone. We stand at the window and see a car in the street; another person, in the interior of the room, tells us that he does not see it. We relate things seen in a dream and learn that other persons did not see them. We find in this way that our description of the physical world differs in some respects from the descriptions of other people. The set of facts we refer to here is the same as expressed by the idea that the immediate world is directly accessible to one person alone.

It is the whole of these facts which is comprehended by the abstractum "ego." We say, "I see the car on the street," and mean by this that the thing "car" is accompanied by other phenomena such as feeling joy in the elegant streamline of the car, or feeling hunger in our stomach; in saying "I" we wish to add that we know well that for other persons the car may be accompanied by rather different phenomena. It is the empirical discovery of the difference between the subjective and the objective world which is expressed by the use of "I." This distinction has entered into the grammar of language, and now language is so impregnated with it that we cannot free ourselves of it and indicate it in almost every phrase. Our preceding description is itself not free from it. We described, some lines previously, the facts leading to the discovery of the ego, and said "*We* stand at the window and see a car . . . . another person . . . . tells *us*. . . . ." Thus in this description we already used the ego-language which we wanted to substantiate. This is, however, no contradiction or vicious circle. We used the usual ego-language only to be more easily understood. We could have given the same description by speaking in a neutral language. The original neutral language does not say "I see" but "There is"; only because we hear that another person answers "There is not" do we retire to the more modest statement "I see."

It is the epistemological transition to the impression

basis which is expressed in this grammatical habit. There is a long line of experience hidden behind this "I." The ego is by no means a directly observed entity; it is an abstractum constructed of concreta and illata as internal elements. Descartes's idea that the ego is the only thing directly known to us and of which we are absolutely sure, is one of the landmarks on the blind alleys of traditional philosophy. It involves mistaking an abstractum for a directly observed entity, mistaking an empirical fact for a priori knowledge, mistaking a product of experience and inferences for the metaphysical basis of the world. Empiricists of all times have rightly opposed it.[14] Let us quote here Lichtenberg, who though he called himself an idealist found the most striking formulation for the empiricist answer to Descartes: "*It thinks*, we ought to say, as we say: *it lightens*. To say *cogito* is already too much, if it is translated by *I think*."[15] The original language is neutral and does not know an ego—this ego is a logical construction.

As the abstractum "ego" is to express an empirical fact, we are free to imagine a world in which there would be no ego. Imagine that all people were connected, according to the salamander operation (§ 27), in such a way that everybody shared the impressions of everybody else. Nobody would then say, I see, or I feel; they would all say, There is. On the other hand, we may obtain the opposite case by dissolving the unity of one person into different egos at different times; if there were no memory, the states of one person at different times would be divided into different persons in the same way that spatially different bodies are

[14] Cf. an interesting summary of the empiricist criticism of the ego-concept given by H. Löwy, *Erkenntnis*, III (1932/33), 324.

[15] "*Es denkt*, sollte man sagen, so wie man sagt: *es blitzt*. Zu sagen *cogito*, ist schon zu viel, sobald man es durch *ich denke* übersetzt" (cf. Lichtenberg's *Vermischte Schriften* [Göttingen, 1844], I, 99).

divided into different persons. The concept of ego then would not have been developed. Voltaire, impressed by the ideas of Hume, knew this when he wrote in his *Dictionnaire philosophique*, in the article "Identité": "Vous n'êtes le même que par le sentiment continu de ce que vous avez été et de ce que vous êtes; vous n'avez le sentiment de votre être passé que par la mémoire: ce n'est donc que la mémoire qui établit l'identité, la mêmeté de votre personne."

We are glad that we may quote older empiricists in the defense of an idea which finds its natural place in modern empiricism as well. We know that our empiricism is not a product of our time alone but finds its place in a long historical development. This has been obscured by the traditional metaphysical way of writing the history of philosophy, which has distorted all objective historical aspects. The prevalence of metaphysicians in the field of history is due, I think, to the fact that they have a special liking for history, whereas empiricists prefer to engage in the analysis of problems. The history of empiricism will have to be rewritten some day by the empiricists themselves.

§ 29. The four bases of epistemological construction

In the foregoing sections we gave an epistemological construction of the world on the concreta basis. We showed first that starting from this basis we construct, by projections, the whole external or physical world; we proceeded then to construct on the same basis, and also by projections, the whole internal or psychical world. The term "psychical," we indicated, is misleading, as the objects constructed are not of a type different from physical objects; they are physiological processes within the human body. The false interpretation of these internal objects as objects of "another sphere," of the "psychical sphere," is a misunderstanding due to an insufficient logical analysis. It is the particular situation of the observer in this case, the necessity of observing or inferring processes within his own body, which causes this misunderstanding so current in traditional philosophy. A correct analysis shows the way to liberation from such misinterpretations.

There is, however, no logical necessity for choosing concreta as the basis for the logical construction of the world. We have already pointed this out several times; we shall proceed now to a systematic survey of the different possible bases of epistemological construction.

The particular position of man as that being who wants to perform the construction suggests a classification which is related to man as point of reference. This idea leads to the distinction of three kinds of bases according to the trichotomy of stimulus, internal process, and reaction:

a) The first is the *concreta basis*, used in the preceding account. It is the *stimulus basis*, i.e., the basis formed by those objects which may become direct stimuli.

b) The second is the *impression basis*. Impressions are internal processes within the human body; thus this positivistic basis is an *internal-process basis*.

c) The third is a *reaction basis*. Among all reactions propositions pronounced by men are the most important; it seems convenient, therefore, to restrict this basis to propositions, i.e., to establish a *proposition basis*.

These bases may be called *anthropocentric*, as they are chosen by reference to man. Before entering into a closer consideration of them, let us add a fourth basis which is not related to man:

d) This fourth is the *atom basis*. By "atom" we may comprehend all those elementary corpuscles such as electrons, protons, photons, which physics has discovered as elements of matter. This basis is not anthropocentric.

The number of possible bases is not restricted. It would be easy to establish other kinds; thus we might consider all physical effects produced on certain physical objects, such as photographic plates, as the basis for a construction of the world. The choice is determined by expediency; the four bases as mentioned constitute the most important types which have been used.

Let us now consider some general relations between these bases. We must first point out a remarkable difference. The bases *a*, *b*, and *d* are similar to one another in so far as they involve objects and may be called *object bases;* the basis *c*, on the contrary is of another logical level, as it is a *sentence basis*. Now the system of knowledge is in itself a human reaction and a sentence reaction; thus the sentence basis, seen in terms of the sentence system of knowledge, is the nearest basis. This leads to some important considerations which we shall develop later.

We shall first consider the object bases. The construction of the world erected upon them is effected by means of projections and reductions. If we use the concreta basis, the illata are constructed by projections and the abstracta by reducibility relations; among the illata are to be placed most of the internal processes of the human body, except those which are accessible to the inner tactile sense. If we use the impression basis, the number of the projections increases, as all concrete physical things are then to be constructed by projection; only certain internal processes are constructed in this case by reducibility relations. The atom basis has the advantage that projections disappear and that the construction is entirely performed in terms of reducibility relations. This may be regarded as the definition of this basis: it is this quality which induces the physicists to use it.

Using mathematical symbolism, we may consider the

basic elements of the epistemological construction as a set of independent variables $x_1 \ldots . x_m$, whereas an entity e constructed on this basis is a function

$$e = f(x_1 \ldots . x_n) \tag{1}$$

where f is a complicated logical function including, in general, projections and reductions, as just described. The introduction of another basis may be considered as the transition to another set of variables $y_1 \ldots . y_m$, by means of functions

$$\left.\begin{array}{l} x_1 = t_1(y_1 \ldots . y_m) \\ \ldots\ldots\ldots\ldots\ldots \\ \ldots\ldots\ldots\ldots\ldots \\ x_m = t_n(y_1 \ldots . y_m) \end{array}\right\} \tag{2}$$

The entity e, in reference to the new variables $y_1 \ldots . y_m$, is expressed then by another function, f', obtained by the introduction of the transformation (2):

$$e = f'(y_1 \ldots . y_m) \tag{3}$$

The functions $t_1 \ldots . t_n$ consist of projections and reductions, as well as f and f'. The occurrence of probability connections within these functions is of great importance; the neglect of this fact constitutes the main fault of the positivistic conception.

The concreta basis has the great advantage of being intuitive; it is the original basis in a psychological and historical sense (cf. § 25). Its disadvantage is its necessitating the concept of subjective existence, introduced by the unavoidable expansion of the concept of immediate existence into a concept encompassing both real things and things seen in a dream, or in a cinema. The impression basis avoids this disadvantage, as there is an objectively existent impression even in the case of a merely subjectively existent thing, such as in the case of a dream. This is why the impression basis is preferred by many epistemologists; it enables us to construct the world by means of the concept of objective existence alone. On the other hand, the disadvantage of the concept of subjective existence must not be overestimated. It is true that this concept may lead philosophers to metaphysical fancies; but this can be avoided if we keep to the fact that every statement concerning subjectively existent things is equivalent to a statement concerning objectively existent impressions. The subjective language, i.e., that part of the immediate language which concerns subjective things, can therefore be translated into an objective language. Subjective objects may thus be compared to the fictive objects of mathematics, such as the "infinitely distant point," or the "imaginary conic section." These words—and this is true for our subjective language also—can be avoided by another mode of speech; but they are very practical because they allow us to use a simple language in cases in which another language would become rather opaque. The impression language has the great disadvantage that it refers mainly to illata and is therefore unintuitive and unpsychological. It has turned out in many branches of modern science that an ideal language does not exist, that the best language for one section of science is not always the best for another. The construction of a universal language, it follows, cannot be freed from certain inconvenient conflicts with the desires of linguistic taste.

It is the advantage of the concept of immediate existence, because of its inclusion of the concept of subjective existence, that it allows us to obtain basic statements of a high degree of certainty; for it is much more certain that there is an immediate thing *A* than that there is an objective thing *A*. The impression basis attains the same advantage by the introduction of the impression of *A*, instead of the thing *A*. But as we saw that the impression can be characterized by us in stimulus language only, the impression of *A* is defined by the immediately existent thing *A*. This is why both modes of speech turn out to be the same.

The atom basis, on the other hand, starts from basic statements of a low certainty, especially when it is not general physical laws which are to be described but individual processes. This is why physicists, for many purposes, cannot renounce an anthropocentric basis. They choose, then, usually the impression basis. This basis corresponds well with physical methods. Imagine a physical instrument which is used as an indicator for other processes; this instrument will record the effects caused in it by the arrival of causal chains started from other phenomena. The instrument thus indicates the last links of causal chains converging toward one physical system and "infers," making use of the causal chains, the more remote phenomena. Impressions may be conceived in a similar way as the last links in causal chains starting from objects throughout the world and converging toward the human body as indicator. Instead of regarding the effects in the interior of the indicator, we may also consider the effects produced on a certain closed surface surrounding the indicator; this comes to the same thing, as all causal chains must pass the surface. The surface may be identical with the surface of the indicator, i.e., with the surface of the human body. Under this conception, impressions are conceived as processes on the surface of the body only; the processes on the retina, the vibrations of the tympanic membrane, and the like are then the physical facts on which all the construction of the world is based. We are thus led once more to our example of the cubical world (§ 14) as an analogy for inferences on the impression basis; the shadows of the birds are causal effects produced by converging causal chains on a surface surrounding the observer.

It must not be forgotten that the impression basis possesses a high degree of certainty only as long as the impression is defined in stimulus language, i.e., as the impression belonging to a certain physical object. If we pass to the internal-process language, the certainty decreases. That there is a two-dimensional optical image of a seen table on the retina is much less certain than that there is a table before me. This is because the direct characterization of the impression is obtained by scientific inferences which presuppose the existence of the concreta. The concreta basis is the original basis in the psychological sense—in the sense that actual thinking starts from it.

The proposition basis needs a discussion apart from the other bases because it is of another level.

It may be objected that sentences are physical entities as well as impressions or the things of the concreta basis; sentences consist of carbon patches, or waves of sound, and are concreta in the same sense as thermometers or manometers or other instruments observed by the physicist. This is true; but the physical things "sentences" are used in a way different from these other things. They are used as symbols, as a co-ordinated set of things, portraying in itself the world as a map portrays a country. The system of knowledge, being composed of sentences, is also a co-ordinated system, copying the world. The sentence basis is for this reason more closely related to knowledge than an object basis; it is of the same nature as the system of knowledge.

This has an advantage. Instead of considering the relations between things or facts, on the sentence basis we may consider relations between sentences. This is the reason

Carnap[16] has insisted on choosing the sentence basis. He maintains that certain relations which are considered as relations between things or facts are originally relations between sentences. Take the relation of implication. We say that "It rains" implies that "the street becomes wet." This is, says Carnap, a relation between sentences. If we consider it as a relation between the corresponding facts, this is a "shifted language"—a language which has left its original basis and assumed another one.

I do not think that this is a question of principle. Whether we should consider implication as a relation between sentences or between facts seems to me a matter of convention. For many purposes it may be convenient to consider it as a relation between sentences—such as the definition of implication as a certain tautological connection between sentences. There is, on the other hand, no difficulty in considering implication as a relation between facts. This corresponds much better to the actual signification of the concepts. Returning to our example, in speaking of an implication we want to express that the fact "raining" is always accompanied by the fact "the becoming wet of the street." It is such a permanent association of facts which we want to express by the word "implication."

It may be objected that the character of necessity belonging to implication cannot be expressed if we define implication as a relation between objects; i.e., that we cannot then distinguish strict implication[17] from general implication. This is true, and certainly an important result of Carnap's investigations. Idealized concepts like "strict necessity," "strict impossibility," "strict implication,"

[16] *Logische Syntax der Sprache* (Vienna, 1934).

[17] The term "strict implication" has been introduced by C. I. Lewis, whereas Carnap usually speaks of "deducibility relation."

concern propositions only and not facts. Empirical observation gives no means of distinguishing between the two propositions: "The fact A strictly implies the fact B" and "The fact A is always accompanied by the fact B"; if we insist, nevertheless, upon a surplus meaning for the first proposition, this is a matter which can only be formulated as a property of the propositions. This property would be, in our example, the tautological connection of the propositions about A and B. But we must bear in mind that the surplus meaning saved by this interpretation is of no relevance for the content of science. Science is to give verifiable information about empirical objects—this aim can be fully attained in object language and needs no addition expressible in proposition language only.

The idea that such relations as implication are relations between sentences has led Carnap to maintain that philosophy is analysis of scientific language. This is, I think, not false, and it may be useful to conceive philosophy under such a definition. We ourselves made use of this conception when we reduced the question of the existence of external things to a question of the meaning of sentences. I should say, nevertheless, that such a definition of philosophy is not in opposition to the view that philosophy is concerned with the analysis of the more general relations holding for the physical world. This second interpretation is valid because scientific language is not arbitrary but constructed in correspondence to facts. There are only some features of language which have no relevance for the object world; among these are the idealized concepts which have been mentioned. There are, however, other features of language which have their origin in certain features of the world. Thus an analysis of language is at the same time an analysis of the structure of the world.

If the second interpretation is forgotten, a danger arises

which may imperil the understanding of philosophical methods. It is the danger that questions of truth-character may be confounded with questions of arbitrary decision. Language contains many arbitrary elements, and analysis of language is synonymous for many people with an analysis of the arbitrary elements of knowledge. This view would involve, however, a profound misunderstanding of the task of philosophy. There are some essential features of language which are not arbitrary but which are due to the correspondence of language with facts; the task of philosophy is to point out these features and to show which features of language reveal structural features of the physical world.

We may give as an example the problem of geometry. Geometry indeed may be conceived as a part of the language of science. This becomes obvious in the recognized relativity of geometry; mathematicians have shown that, if a description of the world is possible in Euclidean geometry, it is possible also in a non-Euclidean geometry, and vice versa. Hence the decision for Euclidean or non-Euclidean geometry may be conceived as a decision for a certain scientific language. In spite of this conventional character of geometry, however, there are certain considerations of truth-character occurring within the problem. It can be shown that the choice of a certain geometry is free only as long as certain definitions, the definitions of co-ordination, have not yet been formulated. After the decision as to these definitions, the question of the geometry of the world becomes an empirical question; i.e., if in different worlds the definitions of co-ordination are settled in the same way, the resulting geometry may be different. Geometrical conventionalism is accordingly a misleading idea; we may regard geometry as conventional only so long as the question of the geometry of the world is not yet put in a sufficiently determinate way. In spite of this, we may keep to the idea that geometry is a feature of scientific language; but it is a feature in which the structure of the physical reality finds its expression.[18]

I should say, therefore, that the sentence basis does not introduce methods different in principle from the methods used in respect to other bases. It is true that every physical observation must be expressed in a sentence if it is to become an element of knowledge, and so it is useful in many cases to start from the sentence and not from the fact. Such a method may also assume the function of furnishing a control in cases in which an object basis may be misleading. But there are other problems in which the sentence basis is misleading.

We juxtaposed the sentence basis to the three kinds of object bases; however, this needs a correction. We may co-ordinate with each of the three object bases a sentence basis, according as the sentence concerns concreta, or impressions, or atoms. Thus the sentence bases repeat the differences of the object bases at another level. Instead of speaking of a particular sentence basis, we had better speak therefore of the sentence form of the basis in question, considering an object basis and the corresponding sentence basis as different forms of the same basis.

The transition from the object basis to the sentence basis is not the transition to another basis and cannot be symbolized by the mathematical transformation (2). It is a transition only to another mode of speech. Which mode of speech is preferable is, however, a matter of expediency and scientific taste.

[18] For the substantiation of these remarks about geometry we may refer to the author's *Philosophie der Raum-Zeit-Lehre*, § 8.

§ 30. The system of weights co-ordinated to the construction of the world

After having exhibited the construction of the world erected on the concreta basis, we proceed now to the question of the distribution of weights within this construction. It is only after adding the co-ordinated system of weights that our construction becomes complete; without this addition the logical construction would lack its internal order as established by the postulate of truth. This is, however, a problem which can be raised only within the probability theory of knowledge, i.e., a theory in which truth has been replaced by the wider concept of probability. For a two-valued system of knowledge, all propositions forming a part of the system of knowledge are equally true; thus there is no internal order among them from the viewpoint of truth. As this obviously contradicts the practice of science as well as of all knowledge of daily life, the possibility of constructing the co-ordinated system of weights may be regarded as a new proof for the superiority of the probability theory of knowledge.

The particular position of the concreta basis is due to the fact that it presents itself in combination with a very high rank of weights. Statements about the concrete things around us, such as houses, furniture, streets, other people, etc., are practically certain, i.e., possess a very high weight which can be considered as certainty for many purposes. The passage from concreta to illata is accompanied by a continuous diminution of weight. That there is a needle pointing to the number 3.4 of a white board is of a very high degree of certainty; that there is a galvanometer before me pointing to 3.4 amperes is less certain (because the term "galvanometer" includes statements concerning further conditions to be fulfilled) but still of a rather high weight; that there is an electrical current of 3.4 amperes is of a lower weight (because this statement presupposes the "working" of the instrument); that the temperature in the electrical oven heated by this current is about 357° C. is of a still lower weight. This chain of inferences is of a type frequently occurring in physics; every physicist knows the order of certainty which we have indicated and will, in case of any failure of his experimental arrangement, start to question the "working "of its parts according to the inverse order of certainty, i.e., beginning with the least certain parts.

The chains of decreasing weight constructed in this way may lead to complicated interconnections. In our example, the chain may lead to a new concretum. It may be that mercury is put into the electric stove; as mercury is evaporated at the temperature of 357° C., this evaporation may be directly observed and so may furnish a control for the chain of inferences. The end of the chain then receives a rather high weight; this reflects upon the middle parts of the chain so that their weight also increases, although remaining a little lower than that of the ends of the chain. Thus a system of interconnections is constructed, and the calculation of the weights becomes a very complicated matter. We shall consider this concatenation of probabilities in the following chapter, where we shall analyze it in a more detailed manner.

The character of the concreta basis, as the point of issue for all these inferences, becomes visible in any case where there is a new and strange experience whose interpretation is not yet determined. Imagine an engineer who discovers a new effect in a vacuum tube, say, a sudden rise of the anodic current when a certain pressure of a specific gas is poured into the tube. At first he will not believe in this physical interpretation of his experience. He will look over his wires, batteries, and screws to ascertain whether the concreta basis of his inferences is unchanged. He will then control his instruments and his set by replacing the tube in question by another tube of known effects; he thus determines whether his concreta basis leads to the usual concrete effects if it is used in a normal way. He connects in this way the observed fact with a wider concreta basis. Whoever takes part in practical work with abstracta or illata—and almost every branch of higher engineering is occupied with such things—will know that this return to the concreta basis is used as the only decisive method of control.

The concreta are the things best known to us; all other knowledge is derived from this primitive knowledge. The question as to the source of this primitive knowledge arises: How do we know the things of the concreta world?

To this we must answer that the concrete things immediately present themselves to us; they appear, they are there—there is no choice left as to whether or not we shall acknowledge them. There is a choice as to pronouncing the statement, and the difference between "truth" and "lie" marks this liberty of speech; but this difference just indicates that there is no liberty left as to knowing about the immediate thing—he who tells a lie knows that his words do not conform with his observations. This is what we call the *peremptory character of immediate things;* the immediate concreta obtrude upon us, whereas we remain passive, receiving information, ready to observe something.

It may be objected that the observed thing may depend on our will; if we want to see an open window, we perhaps turn our heads to the left and see it; if we want to see a closed window, we turn our heads to the right and see it. What is here amenable to our will, however, is not the observed thing but certain conditions which may produce it. These conditions will lead to the desired thing only if there is no disturbance of the physical connections of the thing in question. Someone may have shut the window while I was looking aside; then, if I turn to the left, the open window will not appear, but a closed one will. The phenomenon then will appear contrary to my expectation and will demonstrate the peremptory character of immediate things.

There is, at this stage, no difference as to things which are only subjective and others which are both immediate and objective. The distinction of subjective and objective things is a later correction which we add in order to avoid contradictions. The peremptory character is a quality which is combined with being an immediate thing, independently of its being jointly an objective thing. On the other hand, things which are only objective, not immediate, do not possess this peremptory character.

We may describe our immediate observations in sentences and may imagine a list of report propositions which forms the sentence basis corresponding to our concreta basis. It must not be forgotten, however, that these report propositions must be immediately true, i.e., correspond to the immediately observed objects. We pointed out in our first chapter (§§ 4 and 5) how a proposition can be compared with a fact; we said that it is not a primitive similarity between sentences and facts which occurs here but a rather complicated co-ordination presupposing the rules of language. It is this correspondence with the immediate things which we demand for the report propositions if we insist that they are to be true.

It has been objected that a proposition is not compared to a fact but only to another proposition. If we want to control a certain given proposition a_1 concerning concreta, so this theory argues, we look at the fact, pronounce a second proposition a_2, called a report, and then compare a_1 with a_2. This theory, it seems to me, does not advance our problem. Of course we may intercalate such a second proposition a_2 to which a_1 is to be compared; but then the problem of truth arises for the proposition a_2. We must know that a_2 is true, if this proposition is to control a_1; if we know nothing about the truth of a_2 either, we have now two propositions a_1 and a_2 on an equal level, and, if they contradict each other, we do not know which to prefer.

The answer has been given that the question of preference cannot be decided for two propositions alone; the propositions are incorporated in the whole system of knowledge, and it is by statistical methods, based on the superiority of the greater number, that the choice between a_1 and a_2 is determined. This idea, I think, is only half-right. It is true that the whole system of knowledge intervenes in such a problem and that the truth of a_1 and a_2 is controlled by the weight which these sentences obtain in reference to the whole system of knowledge. But it is not true that the sentences a_1 and a_2 enter into this statistical consideration on equal terms; they have, on the contrary, *initial weights* which determine to a high degree the issue of the calculation. It is this initial weight which includes the problem of the immediate truth of the observation proposition. Whoever refuses to speak of the correspondence of the report proposition to the immediate thing is obliged to speak instead of the initial weight of a report proposition. Thus if a_1 is communicated to us by another person, whereas a_2 is observed by ourselves, the proposition a_2 receives a high initial weight and may defeat the proposition a_1.

Let us consider this procedure by an example. A friend who visited yesterday the mosque of Sultan Ahmet utters the sentence a_1: "The mosque of Sultan Ahmet has four minarets." To control this sentence I walk to this mosque and, looking at it, form the report sentence a_2: "The mosque of Sultan Ahmet has six minarets." Convinced of the truth of my own observation, I will now prefer a_2 and denote a_1 as false. Why do I prefer a_2? Is it because of general statistics concerning mosques? Such statistics on the contrary are against a_2, as all other mosques have only four minarets or less. It is because I myself observe the six minarets that I believe in the sentence a_2. It is the peremptory character of the immediate thing which distinguishes the corresponding proposition a_2 from a_1.

This does not mean that general rules do not intervene in this determination. On the contrary, we make use of them also. In the first place, if we say that our friend made a false report, we presuppose that the two minarets he omitted could not have been constructed in a single day; without the presupposition of such a law about the abilities of architects it might have been true that the mosque had only four minarets yesterday. Second, we make use of general statistics in stating that our own report in such cases is highly reliable. There are other cases in which we prefer the report of another man to our own. Imagine that you stand on the bridge of a liner; the officer on duty points toward the horizon and says, "There is a lighthouse." You look there but do not see it; in spite of this you will prefer to believe that there is a lighthouse, knowing well that in such a case the eyes of an old sailor are more reliable than those of a philosopher. It is this general rule which intervenes here in favor of a proposition contradicting your own report.

This does not contradict, however, the principle of the peremptory character of immediate things. What is shown here is only that we must not infer from this character that the thing is jointly an objective thing. This question is decided only by additional inferences—inferences however which presuppose once more, for other immediate things, their peremptory character. That we may apply, in our example, the empirical rule concerning the superiority of a sailor's eyes is rendered possible only by our acceptance of some other immediate facts: we know by our own observation that the man before us is a sailor, that we are on the sea; we remember that in similar cases when we used our glasses we discovered the lighthouse which the naked eye could not see; we remember also that the captain told us last night that we were to reach the shore next morning, and so forth. Thus it is a set of propositions concerning our own observations and recollections which leads, when combined with certain empirical rules, to the consequence that one of our own observation sentences is not objectively true. If there were no such set distinguished by a high initial weight of truth, the statistical calculation leading to the denial of the objective signification of one of my own observations could not be performed or, rather, its result would be indeterminate, as it would depend on the statistical basis arbitrarily chosen.

To avoid "initial weights" the proposal might be made to consider the whole mass of accessible propositions, all propositions entering on equal terms. Our initial weight, then, would be the result of a preceding statistical calculation carried through on the basis of equal weight of all propositions. Such an idea, however, would lead to a complete arbitrariness of knowledge. Given a certain class of basic propositions, leading to a certain system of knowledge, we may easily enlarge it by addition of arbitrary propositions in such a way that a contrary system of knowledge is determined by it. Thus to get rid of the six minarets of the mosque of Sultan Ahmet we might add a thousand propositions stating that there are only four minarets and other propositions stating that our own eyes are unreliable; we should obtain then a system which led to the consequence that the mosque had only four minarets. If we do not admit such an arbitrary enlargement of the basis of propositions, if we should call this a playing with sentences and not knowledge, we then decide in favor of initial weights; for refusing such arbitrarily added sentences as untrue is to be expressed in our terminology by ascribing to them the initial weight zero. Of course, we do not forbid anyone such play with sentences; what we want to maintain is that such a procedure does not correspond with the actual practice of knowledge. What we call knowledge is based on sentences appearing from the very beginning with a high initial weight, or with the character of immediate truth.

To summarize: The highest initial weights concern the immediately observed concrete objects. They form the center with reference to which the system of weights is erected. Reports of other persons, transmitted orally or in written form, can be considered as true; but before this is done they receive certain weights with reference to what I see and know immediately. All weights so occurring are thus determined as functions of the initial weights; objective truth in the sense of a high probability is a logical function of immediate truth.

We must add, however, a determination concerning time. We observe concreta at any moment in which we are awake or dreaming; but the basis of our world at a determinate moment is only given by the class of immediate concreta we observe just at that moment. It is for that reason that we do not admit reports about formerly seen things as immediate reports but apply to them a control similar to the control of the reports of other persons, based on the immediate concreta world observed at the moment in which the judgment is performed. I find a note that I took this photograph at one three-hundredths of a second and with diaphragm eight; shall I believe this? That depends on what I see now on the film; if there is a person on it, and his silhouette is doubled, the time of the exposure must have been longer. All reports of the past, transmitted by other persons or by myself, appear with an initial weight which is referred to what I know and observe just now. The world of the immediate present, itself bearing the highest weights, is the center of reference for all other weights co-ordinated to propositions about other things; the construction of the world is ordered in such a way that the co-ordinated system of weights has its center in the region of the present concreta. This is what we call the *superiority of the immediate present.*

When wandering through time, we carry the center of weights with us. What is an immediate report at one moment becomes a transmitted report at a later moment; the primary weight it had is changed then into a secondary one derived from other immediate weights. This change of the structure of the system of weights is inevitable. It would be a vain attempt to fix the immediate weights by noting them on paper, with the intention of preserving them for a later time. What we have then, at a later moment, is a note on paper; whether this may be considered as the original immediate weight of the event depends on what we know and observe at the later moment and demands a new determination of its weight derived from the later moment as basis. We can keep the note only but not the event. This is what we call the flowing of time; events emerge, stay one moment in the sphere of the immediate present, and glide along the stream of time into a farther and farther past. We cannot accompany the events, cannot follow them or visit them at their place in time; we remain detained in our position in the immediate present from which, as from the center of the perspective, we see, on the one side, the past events arranged one behind the other and, on the other side, the future events in a corresponding arrangement. It is as if we see the landscape from a moving train, in continuously shifting perspectives, all referred to ourselves as the center. The system of weights on which we erect the world as on a logical trestlework is arranged in the form of projection rays radiating from the immediate present.

§ 31. The transition from immediately observed things to reports

We have shown that the basis of our knowledge is the world of immediate things appearing at one moment, and we added that we may imagine this world's being expressed in a set of propositions, the so-called report propositions. We insisted that these propositions are not arbitrary but that they are bound by the condition of being true reports of what we see. We must inquire now as to the way in which we proceed from things to the sentences.

Let us begin this investigation with a physical example concerning an apparatus possessing abilities similar to those of a "reporter"—a television set. Such a device incorporates a photoelectric cell the entrance of which is directed successively to the different points of the object, following a certain regular zigzag course; the different impulses of light, composed in such a way to form a one-dimensional arrangement, produce within the cell a corresponding series of electric currents the intensity of which varies according to the intensity of the light rays coming from the different points of the object. The series of these electric impacts stands in a correspondence to the object which is to be portrayed; it may be considered as a series of report propositions. It is a true report if the apparatus works correctly, that is, if there is a correspondence, according to the rules defined by the construction of the apparatus, between the two-dimensional object and the one-dimensional set of electric impulses. This example illustrates our physical theory of truth; it shows that a correspondence between objects and a one-dimensional series of symbols is possible. It shows at the same time that the correspondence in question is not a simple similarity; it is a correspondence presupposing complicated rules. We should not recognize the relation between the one-dimensional series of electric impacts, furnished by the transmitter of the television apparatus, and the original object, if we were to observe this series directly, say, heard through a wireless receiver as a series of sounds varying in intensity; we should need complicated intellectual operations to determine whether this linear set of sounds is "true," i.e., whether it corresponds to the original object according to the rules of co-ordination established by the apparatus.

The receiver, standing at the other end of the line of communication, furnishes the control automatically by transforming the one-dimensional series of electric currents into a two-dimensional picture; it transforms the one-dimensional "sentence" consisting of electric currents back into a thing similar to the original and easily compared with it. Thus there is, finally, a transformation of a thing into a picture similar to it; but there is intercalated in the path of transmission a one-dimensional series of "symbols," having no similarity to the object, but carrying in itself, by means of a complicated co-ordination, all the qualities of the object, so that at the end of the transmission process they reappear as features of the picture. We may say that the two television sets, the transmitter and the receiver, must "think" the object before they can produce the picture at the other end of the line of transmission.

It is easy to describe a similar arrangement in which these two electromechanical sets are replaced by men, and in which a so-called genuine thinking occurs. Imagine a man who observes an object and telephones what he sees to another man; this man at the other end of the cable draws the object according to the description. The processes occurring within these two men are of the same type as those occurring in the television set. The first man is the transmitter, the second the receiver; their communication is rendered possible only because they "think" the object, i.e., describe it in language. The description of the object passing through the wire in the form of electric currents stands to the object in a physical correspondence relation of the same type as that occurring between the series of electric currents furnished by the television transmitter and the object copied by it.

In the case of man we do not know sufficiently the mechanism which produces the sentences co-ordinated with the objects; in spite of that fact we may handle this mechanism as satisfactorily as a person without any understanding of higher engineering may handle television apparatus. Such a "handling of an unknown mechanism" is always performed by us when we make reports of the objects observed by us. But the sentences furnished by a man as observer are not of another kind than the sentences furnished by a television transmitter as observer; they are both true because they stand in a correspondence relation to the physical thing they describe.

The television transmitter does not always "work correctly"; there may occur disturbances which result in producing "false" sentences. To control this, the apparatus may show a red lamp which burns as long as the apparatus "works correctly," going out when the apparatus is disturbed. The same thing may happen to the human body as transmitter; the sentences furnished by men may be false, that is, not in correspondence (as established by the rules of language) with the observed facts. This is the case when the observer is lying. The observer himself knows this difference well; he knows whether or not the red lamp of immediate truth is burning during his speech.

The adherents of the sentence language sometimes drop this difference and say, using behavioristic terms, that in the case of a lie there is the subvocally spoken sentence *a* and the vocally spoken sentence *not-a*. However, this is not an exhaustive description of the phenomenon; we must add that the subvocally spoken sentence appears with a high weight, the vocally spoken sentence with the weight zero. Immediate truth is marked by its *evidence;* although this word has been greatly abused in traditional philosophy, we may apply it in the knowledge that it is not to denote an absolute character, that an evident observation proposition may be objectively false, that even a moment later, in a second observation, the proposition may lose its evidence and may be replaced by a contrary proposition showing instead the red lamp of evidence.

In the case of a report given by another person, the difference between immediate truth and a lie is not so easily observed. But a good psychologist may judge, from the behavior of the person and the whole situation, whether he may trust the report. The red lamp of immediate truth is visible for the reporter only; but, if he shows a "normal behavior," other persons may infer that the red lamp is burning for him. The "normal behavior of the reporter" expresses in reaction language what we call the "evidence character" in stimulus language. Reports bearing this reaction criterion may be accepted in the list of report propositions.

The red lamp of the television transmitter is not an absolutely reliable indicator of the proper functioning of the apparatus. The apparatus may be disturbed but only in such a way that the red lamp continues to burn. The same is valid for the red lamp of immediate truth: it may happen that we have the feeling of pronouncing true sentences but that they actually do not correspond to our observations. Of this kind are slips of the tongue and errors in writing a report. They are not lies because the sentence is uttered in good faith, but nonetheless they lead to report propositions lacking immediate truth.

This needs an additional remark. In the case of the television transmitter there are methods to control the breakdown of the apparatus even if the red lamp continues to burn. We must ask whether there are such methods also for the control of immediate truth. There are such methods, but they are not unambiguous. This is because all methods of control concern objective truth; we are not sure, therefore, whether the fault was committed in the utterance of the sentence or whether the immediate thing differed from the objective thing. We mentioned the example of a note about a photographic exposure, stating that it was taken at one three-hundredths of a second, a note which later on is discovered to be false; was the fault committed in writing only or did I subjectively see the number 300 on the shutter in spite of there being objectively indicated another number? There may be a control by the use of recollection images; we may remember that we worked with the number 50 on the shutter and thus shift the fault to the act of noting. This presupposes, however, a definition of co-ordination as to the use of recollection images (cf. § 27). Without such a definition, or an analogous definition for the application of other methods, the question would become a pseudo-problem; but it must not be forgotten that such a definition of co-ordination can be given and that with such a definition the question of the control of immediate truth becomes as reasonable as the analogous question of objective truth.

In general it is only the objective truth of the proposition which we want to control, and thus the question as to its immediate truth is not raised; only in psychological observations does the question of immediate truth arise. This not only occurs in observations of other people where we have to infer from reactions whether a given report is, for the observer, immediately true; we may also observe the phenomenon that our own reports lack immediate truth. This may happen in reports concerning experiences charged with emotion, such as occur in a psychoanalysis; in such cases a certain courage is needed to heed the red lamp of immediate truth.

The control of immediate truth, as well as that of objective truth, is based on the correspondence theory of truth. Just as the electric impacts of the television transmitter are to be in a certain correspondence to the optical object, so the sentences uttered by men are to correspond to the observed things; it does not matter for this comparison whether objective or subjective things are concerned. We have, therefore, in the correspondence postulate a second criterion of immediate truth; this correspondence criterion is to be put beside the evidence criterion, and we may raise the question as to the compatibility of both criteria.

As to the application of the correspondence theory, we refer to our exposition of this theory in § 5. We showed that the sentence *a* and the sentence "*a* is true" concern different facts: *a* concerns a primary fact, say, a steamer's entering the harbor; "*a* is true" concerns a secondary fact, a relation between the steamer's entering the harbor and a set of words. Let us suppose that we consider the primary fact and that the sentence *a* appears as evident. If we want to control this, we have to consider the secondary fact; if then the sentence "*a* is true" appears as evident, it is proved that the evidence criterion and the correspondence criterion for *a* lead to the same result, i.e., do not contradict each other.

The method may be continued; the evident truth of the sentence "*a* is true" may be controlled by the correspondence method because this sentence once more maintains a correspondence between a sentence and a fact. We have to demand, then, that the sentence, "The sentence '*a* is true' is true," occurs as justified by the evidence criterion. If this is the case, the compatibility of both criteria is proved at a higher level.

We see from these considerations that the evidence criterion of truth cannot be dispensed with; it is only shifted to a higher level. The evidence criterion always remains our ultimate criterion; we must look at a fact with our own eyes if we want to control the truth of a sentence, and, if we apply the correspondence definition of truth, this means nothing but directing our eyes to another fact. This is the difference from the case of the television transmitter. To control the function of this apparatus, we need not use the apparatus itself but have other instruments at our disposal. In the case of controlling our own function of reporting, we are obliged to use just the apparatus we want to control; it is as if a television transmitter were to control its own operation by observing itself with its photoelectric cell and transmitting the resulting electric currents. This is why the evidence criterion is superior to the correspondence criterion; the proper functioning of the red lamp of the transmitter is to be controlled by a second transmission process in which once more the red lamp occurs as a criterion of proper operation. However, such procedure is not a vicious circle but a valuable method of control. It might happen that it leads to contradictions; if it does not, this constitutes a confirmation. By confirmation we understand a unilateral control, that is, a control which might prove the falsehood of a method, though it cannot furnish a decisive control of its correctness.

Applying this control to the problem of immediate truth, we may state the fact that in general both criteria lead to the same result—that, if a sentence appears with the evidence criterion, in most cases the control by the correspondence criterion leads to a confirmation. Using the language of our electrotechnical example, we may say that the human body is a good transmitter; it furnishes automatically sentences which may stand control by the correspondence criterion. Thus, although the evidence criterion is indispensable, the correspondence criterion is permissible as well; as a matter of fact, the criteria coincide.

The superiority of the evidence criterion may raise certain doubts as to the interpretation of scientific methods. We found that the feeling of immediate truth is the decisive indication as to the choice of the foundations of the whole system of knowledge. Why do we ascribe such significance to immediate things? If not all of them are objective things, why do we make them the directive factors of scientific thinking, the test of scientific theories, the object of scientific prophecies? Why is it the world of immediate things, and not that of objective things, for which all the labor of scientific work is done?

Our answer to this question is this: It is just this world of immediate things which is relevant for our lives. What makes us gay and happy and unhappy and ill at ease are the immediate things around us—the houses we live in, the food we eat, the books we read, the things our hands create, the friends we talk with; and all of them in the form in which we see them, and hear them, and feel them—in the form of the immediate things which they are for us. We cannot leave this immediate world; we are bound to live in it and must look for its structure and order to find our way through it. There is no question as to whether we should acknowledge it; we are placed in it, and to learn to foresee it and to handle it is the natural task of our life.

Is not this subjectivism? If we content ourselves with such an answer, does it not mean the failure of the attempt to construct knowledge as an objective system, independent of human feelings and subjective determinations?

I do not think that we have to admit this. To state such an interpretation contradicts the feelings with which we meet the world of immediate things. We do not feel immediate things as a creation of our own. We sense them as something imposed on us from outside; they are not dependent on our will; they obtrude upon us, even if it is against our expectations or desires. What we called the "peremptory character" of immediate things is interpreted by us, emotionally, as their objectivity, as their being a world of their own, or at least messengers of such an independent world. This is just the contrary of the emotions associated with the term "subjectivism"; and if the man of science has constantly the feeling of discovering something with an existence of its own, this is just because immediately observed things are not controlled by his will but appear with irrefutable positiveness and stubborn perseverance.

It is true that this statement concerns emotional associations only; we may, however, co-ordinate to it a logical interpretation. The distinction of subjective and objective things is introduced by inferences based on immediate things; if these inferences show, on the one hand, that the immediate thing is not always identical with the objective thing, that those among the immediate things which are merely subjective things are to be considered as a product of both objective things and the human body, these inferences demonstrate, on the other hand, that this product has an objective character also: it denotes a process occurring in the human body. It is this transition to an objective conception of immediate things which is expressed in the transition from the immediate language to an objective language: to speak of impressions, instead of immediate things, means putting an objective thing in the place of an immediate thing. It does not matter in this context that impressions are only inferred and not observed. We know immediate things, and even merely subjective things, such as the objects of a dream, are not empty shades without any connection with the objective world; they indicate in any case internal processes within our own body, and, as our body constitutes a part of the objective world, we know at least something about some small portion of the world. This turn of subjective things into objective things is as justifiable as is the distinction between these two categories: if it is permissible even to speak of some things as merely subjective, it is also permissible to interpret subjective things as indicating objective things of another kind, constituted by processes within the human body.

This conception gives a decisive turn to the problem of the objectivity of knowledge. The idea that all things we observe at least indicate an inner state of our own body must be considered as one of the greatest discoveries which traditional epistemology presents to us; as our body is in a continuous physical connection with other physical things, this discovery unlocks the door of our private world with its individualistic seclusion. There is at least a small domain of the world known to us; we can make it a basis of inferences leading into the remotest parts of the world. It is the idea of projection which opens these windows to the world; we consider the causal chains which project the world to our small observation-stand as indicators of a much wider environment, the structure of which can be retraced if we copy these causal chains by chains of inferences inversely directed.

However, this expansion of our knowledge presupposes the concept of probability. It is only because the methods of probability are at our command that we can construct these chains of inferences. If we had nothing but tautological transformations at our disposal, we could never leave our small platform and would do nothing but repeat in various forms what we there observe. Inferences of probability character, on the contrary, enable us to advance from place to place; they allow us to add to our observations of the personal platform a knowledge about more distant objects. They can do this because they make no pretense of certainty as do tautological transformations; if we advance farther and farther, the degree of certainty decreases—but only because we pay this turnpike toll can we advance.

We have pointed out this function of the concept of probability during all the stages of our inquiry. We showed that the meaningfulness of sentences about the physical world can be kept only if we introduce the probability concept in place of the concept of truth. We demonstrated that under this condition knowledge starting from a given sphere of observation is not bound to this sphere but may advance to things beyond. We applied the same principle to the investigation of the interior world of our own body and showed that it may be inferred with probability from the surrounding world of stimuli and reactions. We could explain the opposition to the physiological interpretation of psychology in terms of a justified antagonism to the identification of statements about stimuli and reactions with statements about inner processes—an antagonism which disappears however if the probability character of these inferential connections is recognized. We showed, finally, that the whole construction of the world is carried by a trestle-work of probability connections which finds its basis in the world of the immediate concreta but leads outward in two opposite directions to the worlds of large and small dimensions. Placed in the middle of the world, we attach to our point of reference, by probability chains, the whole universe.

It is the concept of probability, therefore, which constitutes the nerve of the system of knowledge. As long as this was not recognized—and logicians were particularly blind in this respect—the logical structure of the world was misunderstood and misinterpreted; an error which led to distorted epistemological constructions neither suiting the actual procedure of science nor satisfying the desire to understand knowledge. The concept of probability frees us from these difficulties, being the very instrument of empirical knowledge.

We have used this concept, however, as yet in a naïve way; we have applied it without giving an analysis of its logical structure. It is this task to which we must now turn. It is only from such an analysis that we may expect a final clarification of the nature of knowledge. We may add that this analysis will lead to a surprising result—that it will show the nature of knowledge as being much different from what its usual interpretations claim. In renouncing pretension of the certainty of knowledge, we must be ready to admit a fundamental change in its logical interpretation. But we may leave the exposition of this idea to the following chapter.

CHAPTER V

PROBABILITY AND INDUCTION

CHAPTER V

PROBABILITY AND INDUCTION

§ 32. The two forms of the concept of probability

The concept of probability has been represented in the preceding inquiries by the concept of weight. However, we did not make much use of this equivalence; we dealt with the concept of weight in an independent manner, not regarding the delimitations involved in its presumed equivalence to the concept of probability. We showed that there is such a concept of weight, that knowledge needs it in the sense of a predictional value, and that it is applied in everyday language as well as in scientific propositions—but we did not enter into an analysis of the concept, relying on a layman's understanding of what we meant by the term. We made use of the fact that the handling of a concept may precede an analysis of its structure. We constructed the triplet of predicates—meaning, truth-value, and weight—and found that it is the latter concept to which the others reduce. Truth has turned out to be nothing but a high weight and should not be considered as something other than an idealization approximately valid for certain practical purposes; meaning has been reduced to truth and weight by the verifiability theory—thus we found that the logical place for the concept of weight is at the very foundation of knowledge. It will now be our last task to enter into the analysis of this concept and to prove its equivalence to the probability concept; we may also hope to clarify its functions by their derivation from a concept as definitely determined as the concept of probability.

Turning to this task, we meet the fact that there are two different applications of the concept of probability, only one of which seems to be identical with the concept of weight as introduced by us. At the beginning of our inquiry into the nature of probability, we find ourselves confronted with the necessity of studying this distinction; we have to ask whether we are justified in speaking of only one concept of probability comprising both applications.

There is, first, the sharply determined concept of probability occurring in mathematics, mathematical physics, and all kinds of statistics. This mathematical concept of probability has become the object of a mathematical discipline, the calculus of probability; its qualities have been exactly formulated in mathematical language, and its application has found a detailed analysis in the well-known methods of mathematical statistics. Though this discipline is rather young, it has been developed to a high degree of perfection. This line of development starts with the inquiries of Pascal and Fermat into the theory of games of chance, runs through the fundamental works of Laplace and Gauss, and finds its continuation in our day in the comprehensive work of a great number of mathematicians. Any attempt at a theory of this mathematical concept of probability must start from its mathematical form. Mathematicians, therefore, have endeavored to clarify the foundations of the concept; among modern investigators of this subject, we may mention the names of v. Mises, Tornier, Dörge, Copeland, and Kolmogoroff.

There is, however, a second concept of probability which does not present itself in mathematical form. It is the concept which appears in conversation as "probably," "likely," "presumably"; its application is, however, not confined to colloquial language but is extended to scientific language also, where suppositions and conjectures cannot be avoided. We pronounce scientific statements and scientific theories not with the claim of certainty but in the sense of probable, or highly probable, suppositions. The term "probable" occurring here is not submitted to statistical methods. This logical concept of probability, though indispensable for the construction of knowledge, has not found the exact determination which has been constructed for the mathematical concept. It is true that logicians of all times have considered this concept, from Aristotle to our day; thus the scientific treatment of this concept is much older than that of the mathematical concept which began with the investigations of Pascal and Fermat. But the theory of the logical concept of probability has not been able to attain the same degree of perfection as the theory of the mathematical concept of probability.

It was the great merit of the creators of logistic that they contemplated, from the very beginning, a logic of probability which was to be as exact as the logic of truth. Leibnitz already had demanded "une nouvelle espèce de logique, qui traiterait des degrés de probabilité"; but this demand for a probability logic, like his project of a calculus of the logic of truth, was actualized only in the nineteenth century. After some attempts of De Morgan, it was Boole who developed the first complete calculus of a probability logic, which, in spite of some mistakes later corrected by Peirce, must be regarded as the greatest advance in the history of the logical concept of probability since Aristotle. It was a prophetic sign that the exposition of this probability logic was given in the same work which stands at the basis of the modern development of the logic of truth and falsity: in Boole's *Laws of Thought*. In the subsequent development, the problems of the logic of truth have assumed a much wider extent; probability logic was carried on by isolated authors only, among whom we may mention

Venn and Peirce, and among contemporary writers, Keynes, Lukasiewicz, and Zawirski.

If we regard these two lines of development, the supposition obtrudes that underlying them are two concepts which may show certain similarities and connections, but which are in their logical nature entirely disparate. This *disparity conception* of the two probability concepts has indeed been maintained by a great many authors, in the form either of a conscious or of a tacit assumption. On the other hand, the idea has been maintained that the apparent difference of the two concepts is only superficial, that a closer investigation reveals them as identical, and that only on the basis of an *identity conception* can a deeper understanding of the two probability concepts be obtained. The struggle between these two conceptions occupies to a great extent the philosophic discussion of the probability problem. The issue of this struggle is, indeed, of the greatest importance: since the theory of the mathematical concept of probability has been developed to a satisfactory solution, the identity conception leads to a solution of the philosophic probability problem as a whole, whereas the disparity conception leaves the problem of the logical concept of probability in a rather vague and unsatisfactory state. The latter consequence originates from the fact that a satisfactory theory of this concept, as different from the mathematical one, has not yet been presented.

The disparity conception has its genesis in the fact that the mathematical concept of probability is interpreted in terms of frequency, whereas the logical concept of probability seems to be of a quite different type.

Indeed, the great success of the mathematical theory of probability is due to the fact that it has been developed as a theory of relative frequencies. It is true that the original definition of the degree of probability construed for an application in games of chance was not of the frequency type; Laplace gave the famous formulation of the ratio of the favorable cases to the possible cases, valid under the controversial presupposition of "equally possible" cases. This definition, apparently natural for cases of the type of the die, was abandoned however in all applications of the theory to cases of practical value: statisticians of all kinds did not ask for Laplace's "equally possible" cases but interpreted the numerical value of the probability by the ratio of two frequencies—the frequency of the events of the narrower class considered and the frequency of the events of the wider class to which the probability is referred. The mortality tables of life insurance companies are not based on assumptions of "equally possible" cases; the probabilities occurring there are calculated as fractions the numerator of which is given by the class of the cases of decease, and the denominator of which is determined by the class of the population to which the statistics are referred. The relative frequency thus obtained turned out to be an interpretation of the degree of probability much more useful than that of Laplace. The far-reaching extensions of the mathematical theory, indicated by such concepts as average, dispersion, average error, probability function, and Gaussian law, are due to a definitive abandonment of the Laplace definition and the transition to the frequency theory.

The logical concept of probability, on the contrary, seems to be independent of the frequency interpretation, which for many cases of logical probability appears not at all applicable. We ask for the probability of determinate events, say, of good weather tomorrow, or of Julius Caesar's having been in Britain; there is no statistical concept expressed in the question. It is the problem of the probability of the single case which constitutes the origin of the disparity theory; authors such as Keynes,[1] therefore, base their concept of logical probability essentially on this problem.

Such authors even go so far as to deny a numerical value to logical probability. Keynes has developed the idea that logical probability is merely concerned with establishing an order, a series determined by the concepts of "more probable" and "less probable," in which metrical concepts such as "twice as probable" do not occur. These ideas have been continued by Popper.[2] For these authors, logical probability is a merely topological concept. Other authors do not want to admit such a restriction. Their concept of logical probability is metrical, but not of the frequency type. Logical probability, they say, is concerned with the "rational degree of expectation," a concept which already applies to a single event. It is here that the "equally possible" cases of Laplace find their field of application as furnishing the point of issue for the determination of the degree of expectation which a reasonable being should learn to put in place of feelings as unreasonable as hope and fear.

It will be our first task to enter into a discussion of these questions. We must decide in favor of either the disparity conception or the identity conception of the two forms of the probability concept.

§ 33. Disparity conception or identity conception?

The disparity conception is sometimes substantiated by saying that the mathematical concept of probability states a property of *events*, whereas the logical concept of probability states a property of *propositions*.

[1] J. M. Keynes, *A Treatise on Probability* (London, 1921).

[2] K. Popper, *Logik der Forschung* (Berlin, 1935).

If this were to be the whole content of the disparity conception, we would not attack it; for it is indeed possible to make such a distinction. If we interpret probability as a frequency of events, a probability statement would concern events; if we consider, on the contrary, probability as a generalization of truth, we have to conceive probability as concerning propositions. This is made necessary by the nature of the truth concept; only propositions, not things, can be called true, and our predicate of weight which we want to identify with probability has been introduced also as a predicate of propositions. But, if we apply these reflections to the probability concept, we find that they have only a formal signification and do not touch the central problem of the disparity conception. For, if we interpret the logical concept of probability also by a frequency, both concepts become isomorphic; the mathematical concept is then interpreted by a frequency of events, and the logical concept by a frequency of propositions about events.[3] What the identity conception wants to maintain is just the applicability of the frequency interpretation to the logical concept of probability; thus we see that the thesis of the identity conception is, strictly speaking, an isomorphism of both concepts, or a structural identity. Even from the standpoint of the identity conception we may, therefore, consider the logical concept of probability as a concept of a higher linguistic level: such a distinction involves no difficulties for the theory of probability, as we are in any case obliged to introduce an infinite scale of probabilities of different logical levels (cf. § 41).

There is a second sense in which we have to speak of an identity here. If the frequency interpretation is accepted

[3] This isomorphism follows strictly from the axiomatic construction of the calculus of probability which shows that all laws of probability can be deduced from the frequency interpretation (cf. § 37).

for the logical concept, this concept may be applied also to the statements of mathematical statistics: that is to say, even purely statistical statements admit both the mathematical and the logical conception of probability. A statement about the probability of death from tuberculosis may therefore be interpreted as concerning statistics of cases of tuberculosis, or as concerning statistics of propositions about cases of tuberculosis. On the other hand, the examples given for a logical meaning of the probability concept admit both interpretations as well.

For these reasons we shall use in the following inquiries the term "identity conception" without always mentioning that there is, strictly speaking, a difference of logical levels involved. We use the word "identity" here in the sense of an identity of structure, and our thesis amounts to maintaining *the applicability of the frequency interpretation to all concepts of probability*.

It is this thesis which the disparity conception attacks. We shall have to discuss this question now; if we cannot admit the disparity conception, this is because this conception involves consequences incompatible with the principles of empiricism.

There is, first, the principle of verifiability which cannot be carried through within the disparity conception. If a probability of a single event is admitted, in the sense of a predictional value—i.e., of signifying something concerning future events—there is no possibility of verifying the degree of probability by the observation of the future event in question. For instance we throw a die and expect with the probability 5/6 to obtain a number greater than 1: how can this be verified if we watch one throw only? If the event expected does not occur, this is no refutation of the presumption because the probability 5/6 does not exclude the case of the number 1 occurring. If the event expected occurs, this is not a proof of the correctness of the presumption because the same might happen if the probability were 1/6 only. We might at least say that the occurrence of the event is more compatible with the presumption than is the nonoccurrence. But how distinguish then between different degrees of probability both greater than one-half? If we had said that the probability of the event is not 5/6 but 3/4, how is the verification of this presumption to differ from that of the other?

The difficulty is not removed if we try to restrict probability statements to merely topological statements, eliminating the degree of probability. A statement of the form, "This event is more probable than the other," cannot be verified either, if it concerns a single case. Take two mutually exclusive events which are expected with the respective probabilities 1/6 and 1/4; the second one may happen. Is this a proof that this event was more probable than the other? This cannot be maintained because there is no principle that the more probable event must happen. The topological interpretation of logical probability is accordingly exposed to the same objections as the metrical one.

This analysis makes it obvious that a verification cannot be given if the probability statement is to concern a single case only. The single-case interpretation of the probability statement is not compatible with the verifiability theory of meaning because neither the degree nor the order asserted with the probability statement may be controlled if only one event is considered. One of the elementary principles of empiricism, therefore, is violated with this interpretation.

There is a second difficulty with the disparity conception, occurring if the degree of probability is to be quantitatively determined. We said that, if the frequency interpretation is denied, the concept "equally probable" demands a substantiation by the concept of "equally possible cases," such as in Laplace's formulation. This leads, however, into apriorism. How do we know the "equal possibility"? Laplace's followers are obliged to admit here a kind of "synthetic a priori" judgment; the principle of "insufficient reason" or of "no reason to the contrary" does nothing but maintain this in a disguised form. This becomes obvious if we pass to a frequency statement, which in many cases, such as for dice, is attached to the "equal possibility" statement. How do we know that "equal possibility" implies equal frequency? We are forced to assume a correspondence of reason and reality, such as Kant had postulated.

We shall not enter here into a discussion of this second point, although it has played a great role in older philosophical discussions of the probability problem. We may only mention that the problem of the equally probable cases, such as occur in games of chance, finds a rather simple solution within the mathematical theory; no such presupposition as the principle of "no reason to the contrary" is needed there, and the whole question may be reduced to presuppositions such as occur within the frequency theory of probability.[4] It is obvious that the question would not have assumed so much importance if the frequency theory of probability had been thoroughly accepted. The main point of difference in the discussion between the disparity conception and that of identity is to be sought in the problem of the interpretation of the single case. If it can be shown that the single-case interpretation is avoidable, and that the examples which seem to demand such an interpretation may be submitted to the frequency interpreta-

[4] Cf. the report on this problem in the author's *Wahrscheinlichkeitslehre* (Leiden, 1935), § 65. For all other mathematical details omitted in the following inquiries we may also refer to this book.

tion, the superiority of the identity conception is demonstrated. To carry through this conception thus is identical with showing that the frequency interpretation of probability may always be applied. We shall inquire now whether this is possible.

For the frequency interpretation, a verification of the degree of the probability is possible as soon as the event can be repeated; the frequency observed in a series of events is considered as a control of the degree of probability. This interpretation presupposes, therefore, that the event is not described as an individual happening but as a member of a class; the "repetition" of the event means its inclusion within a class of similar events. In the case of the die, this class is easily constructed; it consists of the different throws of the die. But how construct this class in other examples, such as the case of a historical event of which we speak with a certain probability, or the case of the validity of a scientific theory which we assume not with certainty but only with more or less probability?

It is the view of the adherents of the identity conception that such a class may always be constructed and *must* be constructed if the probability statement is to have meaning. The origin of the single-case interpretation is to be found in the fact that for many cases the construction of the class is not so obviously determined as in the case of the die, or in the fact that ordinary language suppresses a reference to a class, and speaks incorrectly of a single event where a class of events should be considered. If we keep this postulate clearly in mind, we find that the way toward the construction of the corresponding class is indicated in the origin and use of probability statements. Why do we ascribe, say, a high probability to the statement that Napoleon had an attack of illness during the battle of Leipzig, and a smaller probability to the statement that Caspar

Hauser was the son of a prince? It is because chronicles of different ty‸ s report these statements: one type is reliable because its statements, in frequent attempts at control, were confirmed; the other is not reliable because attempts at control frequently led to the refutation of the statement. The transition to the type of the chronicle indicates the class of the frequency interpretation; the probability occurring in the statements about Napoleon's disease, or Caspar Hauser's descent, is to be interpreted as concerning a certain class of historical reports and finds its statistical interpretation in the frequency of confirmations encountered within this class. Or take a statement such as pronounced by a physician, when he considers death in a certain case of tuberculosis highly probable: it is the frequency of death in the class of similar cases which is meant by the degree of probability occurring in the statement.

Although it cannot be denied that the corresponding class is easily determined in such cases, another objection may be raised against our interpretation of the probability statement. It is true, our opponents may argue, that the frequency within such a class is the *origin* of our probability statement; but does the statement *concern* this frequency? The physician will surely base his prediction of the death of his patient on statistics about tuberculosis; but does he mean such statistics when he talks of the determinate patient before him? The patient may be our intimate friend, it is his personal chance of death or life which we want to know; if the answer of the physician concerns a class of similar cases, this may be interesting for a statistician but not for us who want to know the fate of our friend. Perhaps he is just among the small percentage of cases of a happy issue admitted by the statistics; why should we believe in a high probability for his death because statistics about other people furnish such a high percentage?

It is the problem of the *applicability of the frequency interpretation to the single case* which is raised with this objection. This problem plays a great role in the defense of the disparity conception; it is said that the frequency theory may at best furnish a substantiation of the degree of probability but that it cannot be accepted as its interpretation as soon as the probability of a single case is demanded. The objection seems very convincing; I do not think, however, that it holds.

A clarification of the problem can only be given by an analysis of the situation in which we employ probability statements. Why do we ask for the probability of future events, or of past events about which we have no certain knowledge? We might be content with the simple statement that we do not know their truth-value—this attitude would have the advantage of not being exposed to logical criticism. If we do not agree with such a proposal it is because we cannot renounce a decision regarding the event at the moment we are faced with the necessity of acting. Actions demand a decision about unknown events; with our attempt to make this decision as favorable as possible the application of probability statements becomes unavoidable. This reflection determines the way in which the interpretation of probability statements is to be sought: *the meaning of probability statements is to be determined in such a way that our behavior in utilizing them for action can be justified.*

It is in this sense that the frequency interpretation of probability statements can be carried through even if it is the happening or not happening of a single event which is of concern to us. The preference of the more probable event is justified on the frequency interpretation by the *argument in terms of behavior most favorable on the whole:* if we decide to assume the happening of the most probable event, we shall have in the long run the greatest number of successes. Thus although the individual event remains unknown, we do best to believe in the occurrence of the most probable event as determined by the frequency interpretation; in spite of possible failures, this principle will lead us to the best ratio of successes which is attainable.

Some examples may illustrate this point. If we are asked whether or not the side 1 of a die will appear in a throw, it is wiser to decide for "not-1" because, if the experiment is continued, in the long run we will have the greater number of successes. If we want to make an excursion tomorrow, and the weather forecast predicts bad weather, it is better not to go—not because the possibility of good weather is excluded but because, by applying the principle underlying this choice for all our excursions, we shall reduce the cases of bad weather to a minimum. If the physician tells us that our friend will probably die, we decide that it is better to believe him—not because it is impossible that our friend will survive his disease but because such a decision, repeatedly applied in similar cases, will spare us many disappointments.

It might be objected against the frequency interpretation that the principle of the greatest number of successes does not apply in cases in which only one member of the class concerned is ever realized. Throws of a die, or excursions, or cases of disease, are events which often recur; but how about other cases in which there is no repetition? This objection, however, conceives the class to be constructed too narrowly. We may incorporate events of very different types in one class, in the sense of the frequency interpretation, even if the degree of the probability changes from event to event. The calculus of probability has developed a type of probability series with changing probabilities;[5] for this type the frequency interpretation may also be carried through, the frequency being determined by the average of the probabilities which occur. Thus every action of our lives falls within a series of actions. If we consider the numerous actions of daily life which presuppose the probability concept—we press the electric button at the door because there is a probability that the bell will ring, we post a letter because there is a probability that it will arrive at the address indicated, we go to the tram station because there is a certain probability that the tram will come and take us, etc.—these actions combine to form a rather long series in which the frequency interpretation is applicable. The actions of greater importance may be included in another series, including those events which, in a narrower sense, are not repeated. The totality of our actions forms a rather extensive series which, if not submitted to the principle of assuming the most probable event, would lead to a remarkable diminution of successes.

We said that we do best to assume the most probable event; this needs a slight correction for cases in which different degrees of importance are attached to the cases open to our choice. If we are offered a wager in which the stakes are ten to one for the appearance of "number 1" and "some number other than 1" on the face of the die, of course it is more favorable to bet on "number 1." It is, however, again the frequency interpretation which justifies our bet; because of the terms of the wager, we will win more money in the long run by so betting. This case, therefore, is included in our principle of behavior most favorable on the whole. Instead of an amount of money, it may be the importance of an event which assumes a function analogous to that of the winnings in the game. If we expect the arrival of a friend with the prob-

[5] Cf. *ibid.*, § 54.

ability of one-third, we had better go to the station to meet him. In this example, the inconvenience of our friend's arriving without our being at the station is so much greater than the inconvenience of our going there in vain that we prefer having the latter inconvenience in two-thirds of all cases to having the first inconvenience in one-third of all cases. Here it is again the frequency interpretation which justifies our behavior; if the probability of the arrival of our friend is one-hundredth only, we do not go to the station because our inconvenience in going ninety-nine times in vain to the station is greater than his inconvenience in arriving one time without our presence.

These considerations furnish a solution of the problem of the applicability of the frequency interpretation to the single case. Though the meaning of the probability statement is bound to a class of events, the statement is applicable for actions concerned with only a single event. The principle carried through in our foregoing investigations stating that there is as much meaning in propositions as is utilizable for actions, becomes directive once more and leads to a determination as to the meaning of probability statements. We need not introduce a "single-case meaning" of the probability statement; a "class meaning" is sufficient because it suffices to justify the application of probability statements to actions concerned with single events. The disparity conception of the two concepts of probability may be eliminated; the principle of the connection of meaning and action decides in favor of the identity conception.

§ 34. The concept of weight

With these considerations, the superiority of the identity conception is demonstrated in principle. But, to carry through the conception consistently, we are obliged to enter into a further study of the logical position of statements about the single case.

If it is only the frequency of the class which is involved in the probability statement, the individual statement about the single case remains entirely indeterminate so long as it is not yet verified. We expect, say, the appearance of numbers other than 1 on the face of the die with the probability 5/6; what does this mean for the individual throw before us? It does not mean: "It is true that a number other than 1 will appear"; and it does not mean: "It is false that a number other than 1 will appear." We must also add that it does not mean: "It is probable to the degree 5/6 that a number other than 1 will appear"; for the term "probable" concerns the class only, not the individual event. We see that the individual statement is uttered as neither true, nor false, nor probable; in what sense, then is it uttered?

It is, we shall say, a *posit.*[6] We posit the event to which the highest probability belongs as that event which will happen. We do not thereby say that we are convinced of its happening, that the proposition about its happening is true; we only decide to *deal with it* as a true proposition. The word "posit" may express this taking for true, without implying that there is any proof of the truth; the reason why we decide to take the proposition as true is that this decision leads, in repeated applications, to the greatest ratio of successes.

Our posit, however, may have good or bad qualities. If the probability belonging to it is great, it is good; in the contrary case it is bad. The occurrence of considerations of this type is best observed when we consider the gam-

[6] The verb "to posit" has been occasionally used already; I shall venture to use it also as a noun by analogy with the corresponding use of the word "deposit."

bler. The gambler *lays a wager* on the event—this is his posit; he does not thereby ascribe a determinate truth-value to it—he says, however, that positing the event represents for him a determinate value. This value may even be expressed in terms of money—the amount of his stake indicates the value the posit possesses for him. If we analyze the way in which this value is appraised, we find that it contains two components: the first is the amount of money which the man would win if his wager were successful; the second is the probability of success. The arithmetical product of both components may be regarded, in correspondence with concepts in use within the calculus of probability, as the measure of the value the wager has for the gambler.[7] We see that, within this determination of the value, the probability plays the role of a weight; the amount of the possible winnings is weighed in terms of the probability of success, and only the weighed amount determines the value. We may say: *A weight is what a degree of probability becomes if it is applied to a single case.*

This is the logical origin of the term "weight" which we used throughout the preceding inquiries. We understand now why the weight may be interpreted as the predictional value of the sentence; it is the predictional component of the whole value of the sentence which is measured by the weight. With this interpretation, the transition from the frequency theory to the single case is performed. The statement about a single case is not uttered by us with any pretense of its being a true statement; it is uttered in the form of a *posit*, or as we may also say—if we prefer an established word—in the form of a *wager*.[8] The frequency

[7] The occurrence of the arithmetical product here is due to the frequency interpretation. If the wager is frequently repeated, the product mentioned determines the total amount of money falling to the gambler's share.

[8] The German word *Setzung* used in the author's *Wahrscheinlichkeitslehre* has both these significations.

within the corresponding class determines, for the single case, the weight of the posit or wager.

The case of the game may be considered as the paradigm of our position in the face of unknown events. Whenever a prediction is demanded, we face the future like a gambler; we cannot say anything about the truth or falsehood of the event in question—a posit concerning it, however, possesses a determinate weight for us, which may be expressed in a number. A man has an outstanding debt, but he does not know whether his debtor will ever meet his liability. If he wants money today, he may sell his claim for an amount determined by the probability of the debtor's paying; this probability, therefore, is a measure of the present value of the claim in relation to its absolute value and may be called the weight of the claim. We stand in a similar way before every future event, whether it is a job we are expecting to get, the result of a physical experiment, the sun's rising tomorrow, or the next world-war. All our posits concerning these events figure within our list of expectations with a predictional value, a weight, determined by their probability.

Any statement concerning the future is uttered in the sense of a wager. We wager on the sun's rising tomorrow, on there being food to nourish us tomorrow, on the validity of physical laws tomorrow; we are, all of us, gamblers—the man of science, and the business man, and the man who throws dice. Like the latter, we know the weights belonging to our wagers; and, if there is any difference in favor of the scientific gambler, it is only that he does not content himself with weights as low as accepted by the gambler with dice. That is the only difference; we cannot avoid laying wagers because this is the only way to take future events into account.

It is the desire for action which necessitates this gambling. The passive man might sit and wait for what will happen. The active man who wants to determine his own future, to insure his food, and his dwelling, and the life of his family, and the success of his work, is obliged to be a gambler because logic offers him no better way to deal with the future. He may look for the best wagers attainable, i.e., the wagers with the greatest weights,[9] and science will help him to find them. But logic cannot provide him with any guaranty of success.

There remain some objections against our theory of weights which we must now analyze.

The first objection concerns the definition of the weight belonging to the statement of a single event. If probability belongs to a class, its numerical value is determined because for a class of events a frequency of occurrence may be determined. A single event, however, belongs to many classes; which of the classes are we to choose as determining the weight? Suppose a man forty years old has tuberculosis; we want to know the probability of his death. Shall we consider for that purpose the frequency of death within the class of men forty years old, or within the class of tubercular people? And there are, of course, many other classes to which the man belongs.

The answer is, I think, obvious. We take the narrowest class for which we have reliable statistics. In our example, we should take the class of tubercular men of forty years

[9] This remark needs some qualification. The wager with the greatest weight is not always our best wager; if the values, or gains, co-ordinated to events of different probabilities are different in a ratio which exceeds the inverse ratio of the probabilities, the best wager is that on the less probable event (cf. our remark at the end of § 33). Reflections of this type may determine our actions. If we call the wager with the highest weight our best wager, we mean to say "our best wager as far as predictions are concerned." We do not want to take into account in such utterances the value or relevance of the facts concerned. By the use of the word "posit" this ambiguity is avoided, as the term "best posit" is always to signify this narrower meaning.

of age. The narrower the class, the better the determination of the weight. This is to be justified by the frequency interpretation because the number of successful predictions will be the greatest if we choose the narrowest class attainable.[10] A cautious physician will even place the man in question within a narrower class by making an X-ray; he will then use as the weight of the case, the probability of death belonging to a condition of the kind observed on the film. Only when the transition to a new class does not alter the probability may it be neglected; thus the class of persons whose name begins with the same letter as the name of the patient may be put aside.

It is the theory of the classical conception of causality that by including the single case into narrower and narrower classes the probability converges to 1 or to 0, i.e., the occurrence or nonoccurrence of the event is more and more closely determined. This idea has been rejected by quantum mechanics, which maintains that there is a limit to the probability attainable which cannot be exceeded, and that this limit is less than certainty. For practical life, this question has little importance, since we must stop in any case at a class relatively far from the limit. The weight we use, therefore, will not alone be determined by the event but also by the state of our knowledge. This result of our theory seems very natural, as our wagers cannot but depend on the state of our knowledge.[11]

[10] Imagine a class A within which an event of the type B is to be expected with the probability 1/2; if we wager, then, always on B, we get 50 per cent successes. Now imagine the class A split into two classes, A_1 and A_2; in A_1, B has a probability of 1/4, in A_2, B has a probability of 3/4. We shall now lay different wagers according as the event of the type B belongs to A_1, or A_2; in the first case, we wager always on non-B, in the second, on B. We shall then have 75 per cent successes (cf. the author's *Wahrscheinlichkeitslehre*, § 75).

[11] It has been objected against our theory that the probability not only depends on the class but also on the order in which the elements of the class are arranged. The latter is true, but it does not weaken our theory. First, it is an important feature of many statistical phenomena that the frequency structure

Another objection has its origin in the fact that in many cases we are not able to determine a numerical value of the weight. What is the probability that Caesar was in Britain, or that there will be a war next year? It is true that we cannot, for practical reasons, determine this probability; but I do not think that we are to infer from this fact that there is no probability determinable on principle. It is only a matter of the state of scientific knowledge whether there are statistical bases for the prediction of unknown events. We may well imagine methods of counting the success ratio belonging to the reports of historical chronicles of a certain type; and statistical information about wars in relation to sociological conditions is within the domain of scientific possibilities.

It has been argued that in such cases we know only a comparison of probabilities, a "more probable" and "less probable." We might say, perhaps, that this year a war is less probable than last year. This is not false; it is certainly easier to know determinations of a topological order than of a metrical character. The former, however, do not exclude the latter; there is no reason to assume that a metrical determination is impossible. On the contrary, the statistical method shows ways for finding such metrical determinations; it is only a technical matter whether or not we can carry it through.

There are a great many germs of a metrical determination of weights contained in the habits of business and daily life. The habit of betting on almost every thing unknown

---

is independent, to a great extent, of changes in the order. Second, if the order is relevant for the determination of the weight, it is to be included in the prescription; such is the case for contagious diseases (where the probability of an illness occurring depends on the illness or lack of illness of the persons in the environment), or for diseases having a tendency to repeat (where the probability changes if the illness has once occurred), etc. The mathematical theory of probabilities has developed methods for such cases. They do not imply any practical difficulty as to the definition of the weight.

but interesting to us shows that the man of practical life knows more about weights than many philosophers will admit. He has developed a method of instinctive appraisal which may be compared to the appraisal of a good contractor concerning the funds needed in opening a new factory, or to the appraisal by an artillery officer of spatial distances. In both cases, the exact determination by quantitative methods is not excluded; the instinctive appraisal may be, however, a good substitute for it. The man who bets on the outcome of a boxing match, or a horse race, or a scientific investigation, or an explorer's voyage, makes use of such instinctive appraisals of the weight; the height of his stakes indicates the weight appraised. The system of weights underlying all our actions does not possess the elaborate form of the mortality tables of insurance companies; however, it shows metrical features as well as topological ones, and there is good reason to assume that it may be developed to greater exactness by statistical methods.

§ 35. Probability logic

The logical conception considers probability as a generalization of truth; its rules must be developed, therefore, in the form of a logical system. It is this probability logic which we shall now construct.

Let us assume a class of given symbols a, b, c, ... ; they may be propositions, or something similar to them—this may be left open for the present. To every symbol there is co-ordinated a number, the value of which varies between 0 and 1; we call it the probability belonging to the symbol and denote it by

$$P(a)$$

E.g., we may have

$$P(a) = \frac{1}{6}$$

In addition, we have logical symbols at our disposal, such as the signs $\bar{}$ for "not," $\vee$ for "or," a period (.) for "and," $\supset$ for "implies," and $\equiv$ for "is equivalent to." Performing with these signs operations based on the postulate that $P(a)$ is to assume functions similar to those of truth and falsehood in ordinary logic, we obtain a kind of logic which we shall call *probability logic.* As there is no further determination of the term "probability" as it here occurs, probability logic is a formal system, to which we may later give interpretations.

How we are to develop this formal system is not, logically speaking, sufficiently determined. We might invent any system of rules whatever and call it probability logic. This is the reason why the problem of probability logic, and the related problem of a logic of modality, have recently occasioned lively discussion; we have been presented with a great number of ingenious systems, especially in the case of the logic of modality, the advantages of each being emphasized by their various authors. I do not think, however, that the question is to be decided by logical elegance, or by other logical advantages of the proposed systems. The logic we seek is to correspond to the practice of science; and as science has developed the qualities of the probability concept in a very determinate way, there is, practically speaking, no choice left for us. This means that the laws of probability logic must be conformable to the laws of the mathematical calculus of probability; by this relation the structure of probability logic is fully determined. A similar remark applies to the logic of modality; the concepts of "possibility," "necessity," and the like, considered here are used in practice as a topological frame of the probability concept; therefore their structure is to be formulated in systems deducible from the general system of probability logic. The construction of this system by means of a deduction from the rules of the mathematical calculus of probability is, therefore, the fundamental problem of the whole domain. This construction has been carried out; however, we cannot present it in detail but must confine ourselves to a report of the results.[12]

The rules occurring in probability logic resemble the rules of ordinary or alternative logic (the term "two-valued logic" is in use also). However, there are two decisive differences.

The first is that the "truth-value" of the symbols a, b, c, ... , is not bound to the two values "truth" and "falsehood," which may be denoted by 1 and 0, but varies continuously within the whole interval from 0 to 1.

The second is a difference concerning the rules. In the alternative logic, the truth-value of a combination $a \vee b$, or $a \, . \, b$, etc., is determined if the truth-values of a and b are given individually. If we know that a is true and b is true, then we know that $a \, . \, b$ is true; or, if we know that a is true and b is false, we know that $a \vee b$ is true, whereas $a \, . \, b$ in this case would be false. Such a rule does not hold for probability logic. We cannot enter here into a detailed substantiation of this statement; we can only summarize the results obtained.[13] It turns out that the "truth-value" of a combination of a and b is determined only if, in addition to the "truth-values" of a and b separately, the "truth-value" of *one* of the other combinations is given. That is

[12] For a detailed exposition cf. the author's article, "Wahrscheinlichkeitslogik," *Berichte der Berliner Akademie der Wissenschaften* (math.-phys. Kl., 1932); and the author's book *Wahrscheinlichkeitslehre*. As to other publications of the author cf. chap. i, n. 14. For a summary of all contributions to the problem cf. Z. Zawirski, "Über das Verhältnis der mehrwertigen Logik zur Wahrscheinlichkeitslogik," *Studia philosophica*, I (Warsaw, 1935), 407.

[13] Cf. the author's *Wahrscheinlichkeitslehre*, § 73. Instead of making the "truth-value" of a combination dependent on that of another combination, we may introduce as a third independent parameter the "probability of b relative to a" which we write $P(a, b)$. This is the way followed in *Wahrscheinlichkeitslehre*. Both ways amount to the same.

to say: if $P(a)$ and $P(b)$ are given, the value of $P(a \vee b)$, or of $P(a \,.\, b)$, and so on, is not determined; there may be cases in which $P(a)$ and $P(b)$ are, respectively, equal, whereas $P(a \vee b)$ and $P(a \,.\, b)$ are different. If, however, the "truth-value" of *one* of the combinations is known, those of the others may be calculated. We may, e.g., introduce $P(a \,.\, b)$ as a third independent parameter and then determine the "truth-values" of the other combinations as a function of $P(a)$, $P(b)$, and $P(a \,.\, b)$. We have, for instance, the formula

$$P(a \vee b) = P(a) + P(b) - P(a \,.\, b) \tag{1}$$

The necessity of a third parameter for the determination of the "truth-value" of the combinations distinguishes probability logic from alternative logic; it cannot be eliminated but originates from a corresponding indeterminacy in the mathematical calculus. If a and b mean the sides 1 and 2 of the same die, we have

$$P(a \,.\, b) = 0$$

because the sides cannot occur together; the probability of the disjunction then becomes 2/6, which follows from

$$P(a) = P(b) = \tfrac{1}{6}$$

and our formula (1). If on the contrary a and b mean the sides numbered 1 on *two* dice which are thrown together, we have on account of the independence of the throws[14]

$$P(a \,.\, b) = \tfrac{1}{6} \cdot \tfrac{1}{6} = \tfrac{1}{36}$$

and our formula (1) furnishes 11/36 for the probability of the disjunction, in correspondence with well-known rules of the calculus of probability.

[14] We may note that our general formulas are not restricted to the case of independent events but apply to any events whatever.

A similar formula is developed for implication. It is shown to be

$$P(a \supset b) = 1 - P(a) + P(a.b) \tag{2}$$

This case differs from the case of disjunction in so far as two indications, the probability of a and that of the product $a \,.\, b$, suffice to determine the probability of the implication; the latter probability turns out to be independent of the probability of b. We cannot, however, replace the indication of $P(a \,.\, b)$ by that of $P(b)$; this would leave the probability of the implication indeterminate.

For equivalence the equation is

$$P(a \equiv b) = 1 - P(a) - P(b) + 2P(a.b) \tag{3}$$

In this case, the three probabilities $P(a)$, $P(b)$, and $P(a \,.\, b)$ are again needed for the determination of the probability of the term on the left-hand side of the equivalence.

Only for the negation $\bar{a}$ does a formula similar to that of alternative logic obtain:

$$P(\bar{a}) = 1 - P(a) \tag{4}$$

The probability of a suffices to determine that of $\bar{a}$.

These formulas indicate a logical structure more general than that of the two-valued logic; they contain this, however, as a special case. This is easily seen: if we restrict the numerical value of $P(a)$ and $P(b)$ to the numbers 1 and 0, the formulas (1)–(4) furnish automatically the well-known relations of two-valued logic, such as are expressed in the truth-tables of logistic; we have only to add the two-valued truth-table for the logical product $a \,.\, b$, which, in the alternative logic, is not independently given but is a function of $P(a)$ and $P(b)$.[15]

[15] It may be shown that for the special case of truth-values restricted to 0 and 1, the truth-value of the logical product is no longer arbitrary but determined by other rules of probability logic (cf. *Wahrscheinlichkeitslehre*, § 73).

These brief remarks may suffice to indicate the nature of probability logic; this logic turns out to be a generalization of the two-valued logic, since it is applicable in case the arguments form a continuous scale of truth-values. Let us turn now to the question as to the interpretation of the formal system.

If we understand by $a, b, c, \ldots,$ propositions, our probability logic becomes identical with the system of weights which we explained and made use of in our previous inquiries. We shall speak in this interpretation of the *logic of weights.*

However, we may give another interpretation to the symbols $a, b, c, \ldots$. We may understand by the symbol a not one proposition but a series of propositions defined in a special manner. Let us consider a propositional function such as "x_i is a die showing 'side 1' "; the different throws of the die, numbered by the index i, then furnish a series of propositions which are sometimes true, sometimes false, but which are all derived from the same propositional function. We shall speak here of a *propositional series* (a_i). The parentheses are to indicate that we mean the whole series formed by the individual propositions a_i. Or take the propositional function: "x_i is a case of tuberculosis with lethal issue"; it will be sometimes true, sometimes false, if x_i runs through all the domain of tubercular people. If we substitute the symbols $(a_i), (b_i), \ldots,$ in our formulas, we may interpret $P(a_i), P(b_i), \ldots,$ as the limits of the frequencies with which a proposition is true in the propositional series. As to the logical operations, we add the definitions

$$\left.\begin{aligned} [(a_i) \vee (b_i)] &\equiv (a_i \vee b_i) \\ [(a_i) \,.\, (b_i)] &\equiv (a_i \,.\, b_i) \\ [(a_i) \supset (b_i)] &\equiv (a_i \supset b_i) \\ [(a_i) \equiv (b_i)] &\equiv (a_i \equiv b_i) \end{aligned}\right\} \quad (5)$$

which postulate that a logical operation between two propositional series is equivalent to the aggregate of these logical operations between the elements of the propositional series. Our system of formulas then furnishes the laws of probability according to the frequency interpretation. We shall speak, in this case, of the *logic of propositional series.* We see that by these two interpretations the logical conception of probability splits into two subspecies. Probability logic is, formally speaking, a structure of linguistic elements; but we obtain two interpretations of this structure by different interpretations of these elements. If we conceive propositions as elements of this structure, and their weights as their "truth-values," we obtain the *logic of weights.* If we conceive propositional series as elements of the logical structure and the limits of their frequencies as their "truth-values," we obtain the *logic of propositional series.*

We explained above that the identity conception maintains the structural identity of the logical and the mathematical concept of probability; we can proceed now to another form of this thesis. Our logic of weights is the probability logic of propositions; it formulates the rules of what the adherents of the disparity conception would call the logical concept of probability. On the other hand, our logic of propositional series formulates the logical equivalent of the mathematical conception of probability, i.e., a logical system based on the frequency interpretation. What the identity conception maintains is *the identity of both these logical systems;* i.e., first, their structural identity, and, second, the thesis that the concept of weight has no other meaning than can be expressed in frequency statements. The concept of weight is, so to say, a fictional property of propositions which we use as an abbreviation for frequency statements. This amounts to saying that every weight may be conceived, in principle, as determined by a frequency; and that, inversely, every frequency occurring in statistics may be conceived as a weight. If the adherents of the disparity conception will not admit this, it is because in certain cases they see only the weight form of probability and, in others, only the frequency form. There are, however, both forms in every case. In cases such as historical events these philosophers regard only the weight function of probability and do not consider the possibility of constructing a series in which the weight is determined by a frequency. In cases such as the game of dice, or social statistics, these philosophers see only the frequency interpretation of probability and do not observe that the probability thus obtained may be conceived as a weight for every single event of the statistical series. One throw of the die is an individual event in the same sense as Julius Caesar's stay in Britain; both may be incorporated in the logic of weights—but that does not preclude the weight's being determined by a frequency. The statistics necessary for this determination are easily obtained for the die but are very difficult to obtain in the case of Caesar's stay in Britain. We must content ourselves in this case with crude appraisals; but this does not prove an essential disparity of the two cases.

§ 36. The two ways of transforming probability logic into two-valued logic

We must now raise the question as to the transformation of probability logic into alternative logic. By the word "transformation" we do not mean a transition of the type indicated before. The transition by restriction of the domain of variables is a specialization; whether it applies depends on the nature of the variables given. We seek now for a transition which may be carried through for any kind of variables, and which transforms any system of probability logic into two-valued logic.

There are two ways of effecting such a transformation. The first is the method of *division*. In its simplest form, the division is a *dichotomy*. We then cut the scale of probability into two parts by a demarcation value p, for instance, the value $p = \frac{1}{2}$, and make the following definitions:

If $P(a) > p$, a is called true
If $P(a) \leqq p$, a is called false

This procedure furnishes a rather crude classification of probability statements, but it is always applicable and suffices for certain practical purposes.

A more appropriate method of division introduces a three-valued logic. We proceed then by a *trichotomy;* we choose two demarcation values, p_1 and p_2, and define:

If $P(a) \geqq p_2$, a is called true
If $P(a) \leqq p_1$, a is called false
If $p_1 < P(a) < p_2$, a is called indeterminate

If we choose for p_2 a value near 1 and for p_1 a value near 0, the trichotomy method has the advantage that only high probabilities are regarded as truth and only low probabilities as falsehood. As to the intermediate domain of the indeterminate, the procedure corresponds to actual practice: there are many statements which we cannot utilize because their truth-value is unknown. If we drop these indeterminate statements, we may regard the rest as statements of a two-valued logic; in this sense the method of trichotomy also leads to a two-valued logic.

As to the validity of the rules of the two-valued logic for the propositions defined as "true" or "false" by dichotomy or trichotomy, the following remark is to be added.

The operation of negation applies for dichotomy because it leads from one domain into the other on account of the relation expressed in (4), § 35. The same is valid for trichotomy if the limits p_1 and p_2 are situated symmetrically; on account of (4), § 35, the negation of a true statement is then false, and conversely. In the case of the other operations, however, the application of the rules of two-valued logic is permissible only in the sense of an approximation. If, for instance, according to our definitions, a is true, and b is true, we may not always regard the logical product $a \, . \, b$ as also true, for there are certain exceptions. This is the case when $P(a)$ and $P(b)$ are near the limit p_1 or p_2; it may happen then that $P(a \, . \, b)$ is below the limit. Thus if a and b are independent, the value of $P(a \, . \, b)$ is given by the arithmetical product of $P(a)$ and $P(b)$; as these numbers are fractions below 1, their product may lie below the limit, whereas each of them lies above the limit. A similar case is possible for disjunction. In general, if a is false, and b is false, their disjunction $a \vee b$ is false also; it may happen however in our derived logic that in such a case the disjunction is true. This possibility is involved in our formula (1), § 35; if $P(a)$ and $P(b)$ lie below the limit, $P(a \vee b)$ may lie above the limit.

The two-valued logic derived from probability logic by dichotomy is seen to be an approximative logic only. The same is valid for the two-valued or three-valued logic derived by trichotomy. The latter becomes a strict logic only if $p_1 = 0$ and $p_2 = 1$, i.e., if the whole domain between 1 and 0 is called indeterminate. Then exceptions such as those mentioned cannot occur; only in case both a and b are indeterminate is there a certain ambiguity.[16] Such a logic, however, does not apply to physics, as the cases $P(a) = 1$ or $P(a) = 0$ in practice do not occur; there would

[16] Cf. the author's *Wahrscheinlichkeitslehre*, §§ 72 and 74.

be no true or false statements at all in physics if this logic were used. A transformation by division is accordingly bound to remain an approximation.

We turn now to the second method of transformation. It is made possible by the frequency interpretation of probability. We started from a relational system L between elements $a, b, c, \ldots,$

$$L[a, b, c, \ldots]$$

As the "truth-value" of the elements $a, b, c, \ldots,$ varies continuously from 0 to 1, L has the character of a logic with continuous scale and signifies probability logic. We said that we may replace the elements $a, b, c, \ldots,$ by another set of elements $(a_i), (b_i), (c_i), \ldots,$ called propositional series; we have then the system

$$L[(a_i), (b_i), (c_i), \ldots]$$

The truth-value of the elements $(a_i), (b_i), (c_i), \ldots,$ also varies on a continuous scale. Now the propositional series $(a_i), (b_i), \ldots,$ are built up of elements which are propositions of two truth-values only, and the "truth-value" of the propositional series (a_i) may be interpreted as the frequency with which the propositions a_i are true. By this interpretation, the relational system L is transformed into another relational system L_0

$$L_0 [a_i, b_i, c_i, \ldots]$$

We may compare this transition to the introduction of new variables in mathematics. L_0 is nothing but the ordinary two-valued logic.

That is to say: Any statement about propositional series, within the frame of probability logic, may be transformed into a statement within the frame of two-valued logic about the frequency with which propositions in a propositional series are true.

It is upon this transformation that the significance of the frequency interpretation is founded. The frequency interpretation allows us to eliminate the probability logic and to reduce probability statements to statements in the two-valued logic.

This transformation seems to be, in opposition to that by dichotomy or trichotomy, not of an approximative but of a strict character; however, it is so only if two conditions are fulfilled:

1. If the new elements a_i, b_i, ... , are propositions of a strictly two-valued character; and

2. If the statement about the frequency with which propositions are true within a propositional series is of a strictly two-valued character.

These conditions are fulfilled for the purely mathematical calculus of probability; that is the reason why this calculus can be built up entirely within the frame of the two-valued logic. As for the application of this calculus to reality, i.e., to physical statements, these two conditions, however, are not fulfilled; for all statements of empirical science the transition indicated remains nothing but an approximation.

As to the second condition, the difficulty arises from the infinity of the series. A mathematically infinite series is given by a prescription which provides the means of calculating its qualities as far as they are demanded; in particular its relative frequency can be calculated. This is why the second condition offers no difficulties for mathematics. A physically infinite series, however, is known to us only in a determinate initial section; its further continuation is not known to us and remains dependent on the problematical means of induction. A statement about the frequency of a physical series, therefore, cannot be uttered with certainty: this statement is in itself only probable. These reflections lead, as we see, into a theory of probability statements of higher levels; as these considerations involve some additional analyses, we may postpone the discussion of this theory to later sections (§§ 41 and 43). It may be sufficient for the present to state that the second condition cannot be fulfilled for statements of the empirical sciences.

At this point the first condition must be subjected to closer consideration. This condition is not fulfilled in empirical science because there are no propositions which are absolutely verifiable. Such was the result of our previous inquiries; we showed that it is only a schematization when we talk of a strictly true or false proposition. Before the throw of the die, we have only a probability statement about the result of the throw; after the throw we say that we know the result exactly. But, strictly speaking, this is only the transition from a low to a high probability; it is not absolutely certain that there is a die before me on the table showing the side 1. The same is valid for any other proposition whatever; we need not enter again into a discussion of this idea. If we consider the second condition as fulfilled—and for certain purposes this may be practical—this assumption is valid, therefore, only in the sense of a schematization.

We may indicate now what is performed in this schematization. Strictly speaking, the elementary propositions a_i possess for us a weight only; if we replace this weight by truth or falsehood, we perform a transformation by dichotomy or trichotomy. Thus the transformation from L to L_0, by the frequency interpretation, presupposes another tranformation by division concerning the new set of elements.

The frequency interpretation, in introducing the two-valued logic, cannot thereby free us from the approximative character of this logic, even if we take no account of the second condition. This does not involve, however, the view that such a transition is superfluous; on the contrary, it is a procedure with which the degree of the approximation is highly enhanced. That is the reason why this transformation plays a dominant role among the methods of science.

We might try to construct our system of knowledge by giving every proposition an appraised weight; we should then find, however, that in this way we obtain a rather bad system of weights. The actual procedure of science replaces such a direct method by an indirect one, which must be regarded as one of the most perspicacious inventions of science. We begin with a trichotomous transformation, accept the propositions of high and low weight only, and drop the intermediate domain. Applying, then, the frequency interpretation of probability, we construct by counting-processes the weight of the propositions before omitted. This is not the only aim of our calculations; we may even control the weight of the propositions accepted in the beginning and possibly shift them from the supposed place within the scale of weights to a new place. Thus a proposition originally assumed to be true may afterward turn out to be indeterminate or false. This is not a contradiction within statistical method because the alteration of the truth-value of some of the elementary propositions does not, on the whole, greatly influence the frequency. We must constantly insist that what was assumed by appraisal as the weight is confirmed later on by a reduction to the frequency of other statements which are judged by appraisals as well. The original appraisals are thus submitted to a process of dissolution, directed by the frequency interpretation. This process of dissolution leads to a new set of appraisals; the improvement associated with this procedure consists in the fact that every individual appraisal becomes less important, that its possible falsehood influences the whole system less. Thus by concerted action of trichotomy and frequency interpretation we construct a system of weights much more exact than we could obtain by a direct appraisal of the weights.

Within this procedure, the essential function of the frequency interpretation becomes manifest. Although our logic of propositions is not two-valued but of a continuous scale, we need not start knowledge with probability logic. We start with an approximative two-valued logic and develop the continuous scale by means of the frequency interpretation. The same method applies inversely: if a probability statement is given, we verify it by means of the frequency interpretation, in reducing it to statements of an approximative two-valued logic. This approximative logic is better than the original probability logic because it omits the doubtful middle domain of weights. It is the frequency interpretation of probability which makes this reduction possible, for in dissolving weights into frequencies it permits us to confine the direct appraisal of weights to such as are of a high or a low degree. The frequency interpretation frees us from the manipulation of a logical system which is too unhandy for direct use.

We must not forget, however, that the two-valued logic always remains approximative. The system of knowledge is written in the language of probability logic; the two-valued logic is a substitute language suitable only within the frame of an approximation. Any epistemology which overlooks this fact runs the risk of losing itself on the bare heights of an idealization.

§ 37. The aprioristic and the formalistic conception of logic

We must now turn to the question of the origin of the laws of probability logic. This question cannot be separated from the question concerning the origin of logic in general; we must enter, therefore, into an inquiry concerning the nature of logic.

In the history of philosophy there are two interpretations of logic which have played dominant roles, and which have endured to form the main subject matter of discussions on logic in our own day.

For the first interpretation, which we may call the *aprioristic interpretation*, logic is a science with its own authority, whether it is founded in the a priori nature of reason, or in the psychological nature of thought, or in intellectual intuition or evidence—philosophers have provided us with many such phrases, the task of which is to express that we simply have to submit to logic as to a kind of superior command.

Such was the conception of Plato, with visionary insight into ideas superadded; such was the doctrine of most scholastics for whom logic revealed the laws and nature of God; such was the conception of the modern rationalists, Descartes, Leibnitz, and Kant, men who must be considered as the founders of modern apriorism in logic and mathematics. The founders of the modern logic of probability, moreover, were not far removed from such a conception. They discovered that the laws of this logic are as evident as the laws of the older logic; they therefore conceived probability logic as the logic of "rational belief" in events the truth-value of which is not known, and thus as a continuation of a priori logic. Boole conceived his probability logic as an expression of the "laws of thought," choosing this term as the title of his major work; Venn called probability logic "a branch of the general science of evidence," and Keynes, the representative of this conception of probability logic in our day, renews the theory of "rational belief." The dominion of apriorism, therefore, extends even into the ranks of the logisticians.

The second interpretation does not acknowledge logic as a material science and may be called the *formalistic interpretation* of logic. The adherents of this interpretation do not believe in an a priori character of logic. They refuse even to talk of the "laws" of logic, this term suggesting that there is something in the nature of an authority in logic which we have to obey. For them logic is a system of rules which by no means determine the content of science, and which do nothing but furnish a transformation of one proposition into another without any addition to its intension. This conception of logic underlay the struggle of the nominalists in the Middle Ages; it was recognized by those empiricists, such as Hume, who saw the need of an explanation of the claim of necessity by logic; and it was to constitute the basis of the modern development of logistic associated with the names of Hilbert, Russell, Wittgenstein, and Carnap.[17] Wittgenstein gave the important definition of the concept of tautology: A tautology is a formula the truth of which is independent of the truth-values of the elementary propositions contained in it. Logic in this way was defined as the domain of tautological formulas; the view as to the material emptiness of logic found its strict formulation in Wittgenstein's definition.

Carnap added a point of view which was essential for the explanation of the claim of necessity by logic. Logic, he said, in continuation of the ideas of Wittgenstein, deals

[17] It is to be noted here that we use the term "formalistic" in a sense somewhat wider than the sense in use within the discussion of modern logistic, where the formalists are represented by the narrower group centering around Hilbert. The differences between these groups are, however, not essential for our survey.

with language only, not with the objects of language. Language is built up of symbols, the use of which is determined by certain rules. Logical necessity, therefore, is nothing but a relation between symbols due to the rules of language. There is no logical necessity "inherent in things," such as the prophets of all kinds of "ontology" emphasize. The character of necessity is entirely on the side of the symbols; such necessities, however, say nothing about the world because the rules of language are constructed in such a way that they do not restrict the domain of experience.

Logic is accordingly called by Carnap the syntax of language. There are no logical laws of the world, but only syntactical rules of language. What we called a logical fact (§ 1), is to be called in this better terminology a syntactical fact. Instead of speaking of the logical fact that a sentence *b* cannot be deduced from a sentence *a*, it is better to speak of a syntactical fact: the structure of the formulas *a* and *b* is of such a kind that the syntactical relation "deducibility" does not hold between them.

The formalistic conception of logic frees us from all the problems of apriorism, from all questions of a correspondence between mind and reality. It is for this reason the natural logical theory of every empiricism. It does not demand from us any belief in nonempirical laws. What we know about nature is taken from experience; logic does not add anything to the results of experience because logic is empty, is nothing but a system of syntactical rules of language.

Let us ask now whether we may insert probability logic into the formalistic conception of logic. It is obvious that this is, for every variety of empiricism, a basic question. We found that the concept of probability is indispensable for knowledge, that probability logic determines the methods of scientific investigation. If we could not give a formalistic interpretation of probability logic, all efforts of the anti-metaphysicians would have been in vain; in spite of their having overcome the difficulties of the two-valued logic, they would now fail before the concept which forms the very essence of scientific prediction—before the concept of probability. A logistic empiricism would be untenable if we should not succeed in finding a formalistic solution of the probability problem.

There is such a solution. To present it we shall proceed by two steps.

The first step is marked by the frequency interpretation. We showed that probability logic can be transformed into the two-valued logic by the frequency interpretation. Our statement of this transformation needs a supplementary remark. Though it is easily seen that such a transformation is obtained by the frequency interpretation, we do not know immediately whether or not this reduction requires axioms of another kind for which we may have no justification. This question can only be answered by an axiomatical procedure which reduces the mathematical calculus of probability to a system of simple presuppositions sufficient for the deduction of the whole mathematical system; the nature of these axioms has then to be considered.

This procedure has been carried through; it leads to a result of the highest relevance for our problem. It turns out that all theorems of probability reduce to one presupposition only: this is just the frequency interpretation. If probability is interpreted as the limit of the relative frequency in an infinite (or finite) series, all laws of probability reduce to arithmetical laws and, with this, become tautological. The demonstration of this theorem involves some complications, as the theory of mathematical probability refers to a great many types of probability series, the normal series, such as occur in games of chance, being only a special type within this manifold. Even a short indication of this demonstration would unduly lengthen our exposition, so we must content ourselves with a statement of the result.[18]

The consequences of this result for the insertion of probability logic into the formalistic interpretation of logic are obvious: the problem of the justification of the laws of probability logic disappears. These laws are justified, as arithmetical laws, within the formalistic interpretation of mathematics. To see the effect of this result, let us remember the difficulties of the older writers on probability logic. They saw that the laws of probability, although admitted by everybody, cannot be logically deduced from the concept of probability if this concept is to mean something like reasonable expectation, or the chance of the occurrence of a single event; the laws, then, were to be synthetical and a priori. The conception of the "laws of rational belief" which expressed this idea originated from the fact that the deducibility of these laws from the frequency interpretation was not seen. We need no "science of evidence" to prove the laws of probability if we understand by probability the limit of a frequency. On the other hand, this is one of the reasons we must insist on the identity conception of the two probability concepts: if they were disparate, if there were a nonstatistical concept of probability, the justification of its laws by the frequency interpretation could not be given, and the formalistic interpretation of probability logic could not be carried through.[19] We should

[18] This reduction of the calculus of probability to one axiom concerning the existence of a limit of the frequency has been carried through in the author's paper, "Axiomatik der Wahrscheinlichkeitsrechnung," *Mathematische Zeitschrift*, XXXIV (1932), 568. A more detailed exposition has been given in the author's *Wahrscheinlichkeitslehre*.

[19] This fact has not been sufficiently noticed by some modern positivists who have tried to defend the disparity conception against me (cf. my answer to Popper and Carnap in *Erkenntnis*, V [1935], 267).

be driven back into the aprioristic position and should be obliged to believe in laws we cannot justify. It is only the frequency interpretation which frees us from metaphysical assumptions and links the problem of probability with the continuous dissolution of the a priori which marks the development of modern logistic empiricism.

The reduction of the laws of probability to tautologies by the frequency interpretation is only the first step in this direction however. There remains a second step to be taken.

§ 38. The problem of induction

So far we have only spoken of the useful qualities of the frequency interpretation. It also has dangerous qualities.

The frequency interpretation has two functions within the theory of probability. First, a frequency is used as a *substantiation* for the probability statement; it furnishes the reason why we believe in the statement. Second, a frequency is used for the *verification* of the probability statement; that is to say, it is to furnish the meaning of the statement. These two functions are not identical. The observed frequency from which we start is only the basis of the probability inference; we intend to state another frequency which concerns *future observations*. The probability inference proceeds from a known frequency to one unknown; it is from this function that its importance is derived. The probability statement sustains a prediction, and this is why we want it.

It is the problem of induction which appears with this formulation. The theory of probability involves the problem of induction, and a solution of the problem of probability cannot be given without an answer to the question of induction. The connection of both problems is well known; philosophers such as Peirce have expressed the idea that a solution of the problem of induction is to be found in the theory of probability. The inverse relation, however, holds as well. Let us say, cautiously, that the solution of both problems is to be given within the same theory.

In uniting the problem of probability with that of induction, we decide unequivocally in favor of that determination of the degree of probability which mathematicians call the *determination a posteriori*. We refuse to acknowledge any so-called *determination a priori* such as some mathematicians introduce in the theory of the games of chance; on this point we refer to our remarks in § 33, where we mentioned that the so-called determination a priori may be reduced to a determination a posteriori. It is, therefore, the latter procedure which we must now analyze.

By "determination a posteriori" we understand a procedure in which the relative frequency observed statistically is assumed to hold approximately for any future prolongation of the series. Let us express this idea in an exact formulation. We assume a series of events A and $\bar{A}$ (non-A); let n be the number of events, m the number of events of the type A among them. We have then the relative frequency

$$h^n = \frac{m}{n}$$

The assumption of the determination a posteriori may now be expressed:

For any further prolongation of the series as far as s *events* (s > n), *the relative frequency will remain within a small interval around* h^n; *i.e., we assume the relation*

$$h^n - \epsilon \leqq h^s \leqq h^n + \epsilon$$

where ϵ *is a small number.*

This assumption formulates the *principle of induction.* We may add that our formulation states the principle in a form more general than that customary in traditional philosophy. The usual formulation is as follows: induction is the assumption that an event which occurred n times will occur at all following times. It is obvious that this formulation is a special case of our formulation, corresponding to the case $h^n = 1$. We cannot restrict our investigation to this special case because the general case occurs in a great many problems.

The reason for this is to be found in the fact that the theory of probability needs the definition of probability as the limit of the frequency. Our formulation is a necessary condition for the existence of a limit of the frequency near h^n; what is yet to be added is that there is an h^n of the kind postulated for every ϵ however small. If we include this idea in our assumption, our postulate of induction becomes the hypothesis that there is a limit to the relative frequency which does not differ greatly from the observed value.

If we enter now into a closer analysis of this assumption, one thing needs no further demonstration: the formula given is not a tautology. There is indeed no logical necessity that h^s remains within the interval $h^n \pm \epsilon$; we may easily imagine that this does not take place.

The nontautological character of induction has been known a long time; Bacon had already emphasized that it is just this character to which the importance of induction is due. If inductive inference can teach us something new, in opposition to deductive inference, this is because it is not a tautology. This useful quality has, however, become the center of the epistemological difficulties of induction. It was David Hume who first attacked the principle from this side; he pointed out that the apparent constraint of the inductive inference, although submitted to by everybody, could not be justified. We believe in induction; we even cannot get rid of the belief when we know the impossibility of a logical demonstration of the validity of inductive inference; but as logicians we must admit that this belief is a deception—such is the result of Hume's criticism. We may summarize his objections in two statements:

1. We have no logical demonstration for the validity of inductive inference.

2. There is no demonstration a posteriori for the inductive inference; any such demonstration would presuppose the very principle which it is to demonstrate.

These two pillars of Hume's criticism of the principle of induction have stood unshaken for two centuries, and I think they will stand as long as there is a scientific philosophy.

In spite of the deep impression Hume's discovery made on his contemporaries, its relevance was not sufficiently noticed in the subsequent intellectual development. I do not refer here to the speculative metaphysicians which the nineteenth century presented to us so copiously, especially in Germany; we need not be surprised that they did not pay any attention to objections which so soberly demonstrated the limitations of human reason. But empiricists, and even mathematical logicians, were no better in this respect. It is astonishing to see how clear-minded logicians, like John Stuart Mill, or Whewell, or Boole, or Venn, in writing about the problem of induction, disregarded the bearing of Hume's objections; they did not realize that any logic of science remains a failure so long as we have no theory of induction which is not exposed to Hume's criticism. It was without doubt their logical apriorism which prevented them from admitting the unsatisfactory character of their own theories of induction. But it remains incomprehensible that their empiricist principles did not lead them to attribute a higher weight to Hume's criticism.

It has been with the rise of the formalistic interpretation of logic in the last few decades that the full weight of Hume's objections has been once more realized. The demands for logical rigor have increased, and the blank in the chain of scientific inferences, indicated by Hume, could no longer be overlooked. The attempt made by modern positivists to establish knowledge as a system of absolute certainty found an insurmountable barrier in the problem of induction. In this situation an expedient has been proposed which cannot be regarded otherwise than as an act of despair.

The remedy was sought in the principle of retrogression. We remember the role this principle played in the truth theory of the meaning of indirect sentences (§ 7); positivists who had already tried to carry through the principle within this domain now made the attempt to apply it to the solution of the problem of induction. They asked: Under what conditions do we apply the inductive principle in order to infer a new statement? They gave the true answer: We apply it when a number of observations is made which concern events of a homogeneous type and which furnish a frequency h^n for a determinate kind of events among them. What is inferred from this? You suppose, they said, that you are able to infer from this a similar future prolongation of the series; but, according to the principle of retrogression, this "prediction of the future" cannot have a meaning which is more than a repetition of the premises of the inference—it means nothing but stating, "There *was* a series of observations of such and such kind." The meaning of a statement about the future is a statement about the past—this is what furnishes the application of the principle of retrogression to inductive inference.

I do not think that such reasoning would convince any

sound intellect. Far from considering it as an analysis of science, I should regard such an interpretation of induction rather as an act of intellectual suicide. The discrepancy between actual thinking and the epistemological result so obtained is too obvious. The only thing to be inferred from this demonstration is that the principle of retrogression does not hold if we want to keep our epistemological construction in correspondence with the actual procedure of science. We know pretty well that science wants to foresee the future; and, if anybody tells us that "foreseeing the future" means "reporting the past," we can only answer that epistemology should be something other than a play with words.

It is the postulate of utilizability which excludes the interpretation of the inductive inference in terms of the principle of retrogression. If scientific statements are to be utilizable for actions, they must pass beyond the statements on which they are based; they must concern future events and not those of the past alone. To prepare for action presupposes—besides a volitional decision concerning the aim of the action—some knowledge about the future. If we were to give a correct form to the reasoning described, it would amount to maintaining that there is no demonstrable knowledge about the future. This was surely the idea of Hume. Instead of any pseudo-solution of the problem of induction, we should then simply confine ourselves to the repetition of Hume's result and admit that the postulate of utilizability cannot be satisfied. The truth theory of meaning leads to a Humean skepticism—this is what follows from the course of the argument.

It was the intention of modern positivism to restore knowledge to absolute certainty; what was proposed with the formalistic interpretation of logic was nothing other than a resumption of the program of Descartes. The great founder of rationalism wanted to reject all knowledge which could not be considered as absolutely reliable; it was the same principle which led modern logicians to a denial of a priori principles. It is true that this principle led Descartes himself to apriorism; but this difference may be considered as a difference in the stage of historical development—his rationalistic apriorism was to perform the same function of sweeping away all untenable scientific claims as was intended by the later struggle against a priori principles. The refusal to admit any kind of material logic—i.e., any logic furnishing information about some "matter"—springs from the Cartesian source: It is the ineradicable desire of absolutely certain knowledge which stands behind both the rationalism of Descartes and the logicism of positivists.

The answer given to Descartes by Hume holds as well for modern positivism. There is no certainty in any knowledge about the world because knowledge of the world involves predictions of the future. The ideal of absolutely certain knowledge leads into skepticism—it is preferable to admit this than to indulge in reveries about a priori knowledge. Only a lack of intellectual radicalism could prevent the rationalists from seeing this; modern positivists should have the courage to draw this skeptical conclusion, to trace the ideal of absolute certainty to its inescapable implications.

However, instead of such a strict disavowal of the predictive aim of science, there is in modern positivism a tendency to evade this alternative and to underrate the relevance of Hume's skeptical objections. It is true that Hume himself is not guiltless in this respect. He is not ready to realize the tragic consequences of his criticism; his theory of inductive belief as a habit—which surely cannot be called a solution of the problem—is put forward with the intention of veiling the gap pointed out by him between experience and prediction. He is not alarmed by his discovery; he does not realize that, if there is no escape from the dilemma pointed out by him, science might as well not be continued—there is no use for a system of predictions if it is nothing but a ridiculous self-delusion. There are modern positivists who do not realize this either. They talk about the formation of scientific theories, but they do not see that, if there is no justification for the inductive inference, the working procedure of science sinks to the level of a game and can no longer be justified by the applicability of its results for the purpose of actions. It was the intention of Kant's synthetic a priori to secure this working procedure against Hume's doubts; we know today that Kant's attempt at rescue failed. We owe this critical result to the establishment of the formalistic conception of logic. If, however, we should not be able to find an answer to Hume's objections within the frame of logistic formalism, we ought to admit frankly that the antimetaphysical version of philosophy led to the renunciation of any justification of the predictive methods of science—led to a definitive failure of scientific philosophy.

Inductive inference cannot be dispensed with because we need it for the purpose of action. To deem the inductive assumption unworthy of the assent of a philosopher, to keep a distinguished reserve, and to meet with a condescending smile the attempts of other people to bridge the gap between experience and prediction is cheap self-deceit; at the very moment when the apostles of such a higher philosophy leave the field of theoretical discussion and pass to the simplest actions of daily life, they follow the inductive principle as surely as does every earth-bound mind. In any action there are various means to the realization of our aim; we have to make a choice, and we decide in accordance with the inductive principle. Although there is no means which will produce with certainty the desired effect, we do not leave the choice to chance but prefer the means indicated by the principle of induction. If we sit at the wheel of a car and want to turn the car to the right, why do we turn the wheel to the right? There is no certainty that the car will follow the wheel; there are indeed cars which do not always so behave. Such cases are fortunately exceptions. But if we should not regard the inductive prescription and consider the effect of a turn of the wheel as entirely unknown to us, we might turn it to the left as well. I do not say this to suggest such an attempt; the effects of skeptical philosophy applied in motor traffic would be rather unpleasant. But I should say a philosopher who is to put aside his principles any time he steers a motorcar is a bad philosopher.

It is no justification of inductive belief to show that it is a habit. It *is* a habit; but the question is whether it is a good habit, where "good" is to mean "useful for the purpose of actions directed to future events." If a person tells me that Socrates is a man, and that all men are mortal, I have the habit of believing that Socrates is mortal. I know, however, that this is a good habit. If anyone had the habit of believing in such a case that Socrates is not mortal, we could demonstrate to him that this was a bad habit. The analogous question must be raised for inductive inference. If we should not be able to demonstrate that it is a good habit, we should either cease using it or admit frankly that our philosophy is a failure.

Science proceeds by induction and not by tautological transformations of reports. Bacon is right about Aristotle; but the *novum organon* needs a justification as good as that of the *organon*. Hume's criticism was the heaviest blow against empiricism; if we do not want to dupe our consciousness of this by means of the narcotic drug of aprioristic rationalism, or the soporific of skepticism, we must find a defense for the inductive inference which holds as well as does the formalistic justification of deductive logic.

§ 39. The justification of the principle of induction

We shall now begin to give the justification of induction which Hume thought impossible. In the pursuit of this inquiry, let us ask first what has been proved, strictly speaking, by Hume's objections.

Hume started with the assumption that a justification of inductive inference is only given if we can show that inductive inference must lead to success. In other words, Hume believed that any justified application of the inductive inference presupposes a demonstration that the conclusion is true. It is this assumption on which Hume's criticism is based. His two objections directly concern only the question of the truth of the conclusion; they prove that the truth of the conclusion cannot be demonstrated. The two objections, therefore, are valid only in so far as the Humean assumption is valid. It is this question to which we must turn: Is it necessary, for the justification of inductive inference, to show that its conclusion is true?

A rather simple analysis shows us that this assumption does not hold. Of course, if we were able to prove the truth of the conclusion, inductive inference would be justified; but the converse does not hold: a justification of the inductive inference does not imply a proof of the truth of the conclusion. The proof of the truth of the conclusion is only a sufficient condition for the justification of induction, not a necessary condition.

The inductive inference is a procedure which is to furnish us the best assumption concerning the future. If we do not know the truth about the future, there may be nonetheless a best assumption about it, i.e., a best assumption relative to what we know. We must ask whether such a characterization may be given for the principle of induction. If this turns out to be possible, the principle of induction will be justified.

An example will show the logical structure of our reasoning. A man may be suffering from a grave disease; the physician tells us: "I do not know whether an operation will save the man, but if there *is* any remedy, it is an operation." In such a case, the operation would be justified. Of course, it would be better to know that the operation will save the man; but, if we do not know this, the knowledge formulated in the statement of the physician is a sufficient justification. If we cannot realize the sufficient conditions of success, we shall at least realize the necessary conditions. If we were able to show that the inductive inference is a necessary condition of success, it would be justified; such a proof would satisfy any demands which may be raised about the justification of induction.

Now obviously there is a great difference between our example and induction. The reasoning of the physician presupposes inductions; his knowledge about an operation as the only possible means of saving a life is based on inductive generalizations, just as are all other statements of empirical character. But we wanted only to illustrate the logical structure of our reasoning. If we want to regard such a reasoning as a justification of the principle of induction, the character of induction as a necessary condition of success must be demonstrated in a way which does not presuppose induction. Such a proof, however, can be given.

If we want to construct this proof, we must begin with a determination of the aim of induction. It is usually said that we perform inductions with the aim of foreseeing the future. This determination is vague; let us replace it by a formulation more precise in character:

The aim of induction is to find series of events whose frequency of occurrence converges toward a limit.

We choose this formulation because we found that we need probabilities and that a probability is to be defined as the limit of a frequency; thus our determination of the aim of induction is given in such a way that it enables us to apply probability methods. If we compare this determination of the aim of induction with determinations usually given, it turns out to be not a confinement to a narrower aim but an expansion. What we usually call "foreseeing the future" is included in our formulation as a special case; the case of knowing with certainty for every event A the event B following it would correspond in our formulation to a case where the limit of the frequency is of the numerical value 1. Hume thought of this case only. Thus our inquiry differs from that of Hume in so far as it conceives the aim of induction in a generalized form. But we do not omit any possible applications if we determine the principle of induction as the means of obtaining the limit of a frequency. If we have limits of frequency, we have all we want, including the case considered by Hume; we have then the laws of nature in their most general form, including both statistical and so-called causal laws—the latter being nothing but a special case of statistical laws, corresponding to the numerical value 1 of the limit of the frequency. We are entitled, therefore, to consider the determination of the limit of a frequency as the aim of the inductive inference.

Now it is obvious that we have no guaranty that this aim is at all attainable. The world may be so disorderly that it is impossible for us to construct series with a limit. Let us introduce the term "predictable" for a world which is sufficiently ordered to enable us to construct series with a limit. We must admit, then, that we do not know whether the world is predictable.

But, if the world is predictable, let us ask what the logical function of the principle of induction will be. For this purpose, we must consider the definition of limit. The frequency h^n has a limit at p, if for any given ϵ there is an n such that h^n is within $p \pm \epsilon$ and remains within this interval for all the rest of the series. Comparing our formulation of the principle of induction (§ 38) with this, we may infer from the definition of the limit that, if there is a limit, there is an element of the series from which the principle of induction leads to the true value of the limit. In this sense the principle of induction is a necessary condition for the determination of a limit.

It is true that, if we are faced with the value h^n for the frequency furnished by our statistics, we do not know whether this n is sufficiently large to be identical with, or beyond, the n of the "place of convergence" for ϵ. It may be that our n is not yet large enough, that after n there will be a deviation greater than ϵ from p. To this we may answer: We are not bound to stay at h^n; we may continue our procedure and shall always consider the last h^n obtained as our best value. This procedure must at sometime lead to the true value p, if there is a limit at all; the applicability of this procedure, as a whole, is a necessary condition of the existence of a limit at p.

To understand this, let us imagine a principle of a contrary sort. Imagine a man who, if h^n is reached, always makes the assumption that the limit of the frequency is at $h^n + a$, where a is a fixed constant. If this man continues his procedure for increasing n, he is sure to miss the limit; this procedure must at sometime become false, if there is a limit at all.

We have found now a better formulation of the necessary condition. We must not consider the individual assumption for an individual h^n; we must take account of the procedure of continued assumptions of the inductive type. The applicability of this procedure is the necessary condition sought.

If, however, it is only the whole procedure which constitutes the necessary condition, how may we apply this idea to the individual case which stands before us? We want to know whether the individual h^n observed by us differs less than ϵ from the limit of the convergence; this neither can be guaranteed nor can it be called a necessary condition of the existence of a limit. So what does our idea of the necessary condition imply for the individual case? It seems that for our individual case the idea turns out to be without any application.

This difficulty corresponds in a certain sense to the difficulty we found in the application of the frequency interpretation to the single case. It is to be eliminated by the introduction of a concept already used for the other problem: the concept of posit.

If we observe a frequency h^n and assume it to be the approximate value of the limit, this assumption is not maintained in the form of a true statement; it is a posit such as we perform in a wager. We posit h^n as the value of the limit, i.e., we wager on h^n, just as we wager on the side of a die. We know that h^n is our best wager, therefore we posit it. There is, however, a difference as to the type of posit occurring here and in the throw of the die.

In the case of the die, we know the weight belonging to the posit: it is given by the degree of probability. If we posit the case "side other than that numbered 1," the weight of this posit is 5/6. We speak in this case of a posit with appraised weight, or, in short, of an *appraised posit.*

In the case of our positing h^n, we do not know its weight. We call it, therefore, a *blind posit.* We know it is our best posit, but we do not know how good it is. Perhaps, although our best, it is a rather bad one.

The blind posit, however, may be corrected. By continuing our series, we obtain new values h^n; we always choose the last h^n. Thus the blind posit is of an approximative type; we know that the method of making and correcting such posits must in time lead to success, in case there is a limit of the frequency. It is this idea which furnishes the justification of the blind posit. The procedure described may be called the *method of anticipation;* in choosing h^n as our posit, we anticipate the case where n is the "place of convergence." It may be that by this anticipation we obtain a false value; we know, however, that a continued anticipation must lead to the true value, if there is a limit at all.

An objection may arise here. It is true that the principle of induction has the quality of leading to the limit, if there is a limit. But is it the only principle with such a property? There might be other methods which also would indicate to us the value of the limit.

Indeed, there might be. There might be even better methods, i.e., methods giving us the right value p of the limit, or at least a value better than ours, at a point in the series where h^n is still rather far from p. Imagine a clairvoyant who is able to foretell the value p of the limit in such an early stage of the series; of course we should be very glad to have such a man at our disposal. We may, however, without knowing anything about the predictions of the clairvoyant, make two general statements concerning them: (1) The indications of the clairvoyant can differ, if they are true, only in the beginning of the series, from those given by the inductive principle. In the end there must be an asymptotical convergence between the indications of the clairvoyant and those of the inductive principle. This follows from the definition of the limit. (2) The clairvoyant might be an imposter; his prophecies might be false and never lead to the true value p of the limit.

The second statement contains the reason why we cannot admit clairvoyance without control. How gain such control? It is obvious that the control is to consist in an application of the inductive principle: we demand the forecast of the clairvoyant and compare it with later observations; if then there is a good correspondence between the forecasts and the observations, we shall infer, by induction, that the man's prophecies will also be true in the future. Thus it is the principle of induction which is to decide whether the man is a good clairvoyant. This distinctive position of the principle of induction is due to the fact that we know about its function of finally leading to the true value of the limit, whereas we know nothing about the clairvoyant.

These considerations lead us to add a correction to our formulations. There are, of course, many necessary conditions for the existence of a limit; that one which we are to use however must be such that its character of being necessary must be known to us. This is why we must prefer the inductive principle to the indications of the clairvoyant and control the latter by the former: we control the unknown method by a known one.

Hence we must continue our analysis by restricting the search for other methods to those about which we may know that they must lead to the true value of the limit. Now it is easily seen not only that the inductive principle will lead to success but also that every method will do the same if it determines as our wager the value

$$h^n + c_n$$

where c_n is a number which is a function of n, or also of h^n, but bound to the condition

$$\lim_{n=\infty} c_n = 0$$

Because of this additional condition, the method must lead to the true value p of the limit; this condition indicates that all such methods, including the inductive principle, must converge asymptotically. The inductive principle is the special case where

$$c_n = 0$$

for all values of n.

Now it is obvious that a system of wagers of the more general type may have advantages. The "correction" c_n may be determined in such a way that the resulting wager furnishes even at an early stage of the series a good approximation of the limit p. The prophecies of a good clairvoyant would be of this type. On the other hand, it may happen also that c_n is badly determined, i.e., that the convergence is delayed by the correction. If the term c_n is arbitrarily formulated, we know nothing about the two possibilities. The value $c_n = 0$—i.e., the inductive principle—is therefore the value of the smallest risk; any other determination may worsen the convergence. This is a practical reason for preferring the inductive principle.

These considerations lead, however, to a more precise formulation of the logical structure of the inductive inference. We must say that, if there is any method which leads to the limit of the frequency, the inductive principle will do the same; if there is a limit of the frequency, the inductive principle is a sufficient condition to find it. If we omit now the premise that there is a limit of the frequency, we cannot say that the inductive principle is the necessary condition of finding it because there are other methods using a correction c_n. There is a set of equivalent conditions such that the choice of one of the members of the set is necessary if we want to find the limit; and, if there is a limit, each of the members of the set is an appropriate method for finding it. We may say, therefore, that the *applicability* of the inductive principle is a necessary condition of the existence of a limit of the frequency.

The decision in favor of the inductive principle among the members of the set of equivalent means may be substantiated by pointing out its quality of embodying the smallest risk; after all, this decision is not of a great relevance, as all these methods must lead to the same value of the limit if they are sufficiently continued. It must not be forgotten, however, that the method of clairvoyance is not, without further ado, a member of the set because we do not know whether the correction c_n occurring here is submitted to the condition of convergence to zero. This must be proved first, and it can only be proved by using the inductive principle, viz., a method known to be a member of the set: this is why clairvoyance, in spite of all occult pretensions, is to be submitted to the control of scientific methods, i.e., by the principle of induction.

It is in the analysis expounded that we see the solution of Hume's problem.[20] Hume demanded too much when he wanted for a justification of the inductive inference a proof that its conclusion is true. What his objections demonstrate is only that such a proof cannot be given. We do not perform, however, an inductive inference with the pretension of obtaining a true statement. What we obtain is a

[20] This theory of induction was first published by the author in *Erkenntnis*, III (1933), 421–25. A more detailed exposition was given in the author's *Wahrscheinlichkeitslehre*, § 80.

wager; and it is the best wager we can lay because it corresponds to a procedure the applicability of which is the necessary condition of the possibility of predictions. To fulfil the conditions sufficient for the attainment of true predictions does not lie in our power; let us be glad that we are able to fulfil at least the conditions necessary for the realization of this intrinsic aim of science.

§ 40. Two objections against our justification of induction

Our analysis of the problem of induction is based on our definition of the aim of induction as the evaluation of a limit of the frequency. Certain objections may be raised as to this statement of the aim of induction.

The first objection is based on the idea that our formulation demands too much, that the postulate of the existence of the limit of the frequency is too strong a postulate. It is argued that the world might be predictable even if there are no limits of frequencies, that our definition of predictability would restrict this concept too narrowly, excluding other types of structure which might perhaps be accessible to predictions without involving series of events with limits of their frequencies. Applied to our theory of induction, this objection would shake the cogency of our justification; by keeping strictly to the principle of induction, the man of science might exclude other possibilities of foreseeing the future which might work even if the inductive inference should fail.[21]

To this we must reply that our postulate does not demand the existence of a limit of the frequency for all series of events. It is sufficient if there is a certain number of series of this kind; by means of these we should then be

[21] This objection has been raised by P. Hertz, *Erkenntnis*, VI (1936), 25; cf. also my answer, *ibid.*, p. 32.

able to determine the other series. We may imagine series which oscillate between two numerical values of the frequency; it can be shown that the description of series of this type is reducible to the indication of determinable subseries having a limit of the frequency. Let us introduce the term *reducible series* for series which are reducible to other series having a limit of their frequency; our definition of predictability then states only that the world is constituted by reducible series. The inductive procedure, the method of anticipation and later correction, will lead automatically to distinguishing series having a limit from other series and to the description of these others by means of the series having a limit. We cannot enter here into the mathematical details of this problem; for an elaboration of this we must refer to another publication.[22]

To elude our defense, the objection might be continued by the construction of a world in which there is no series having a limit. In such a world, so our adversary might argue, there might be a clairvoyant who knows every event of a series individually, who could foretell precisely what would happen from event to event—is not this "foreseeing the future" without having a limit of a frequency at one's disposal?

We cannot admit this. Let us call C the case in which the prediction of the clairvoyant corresponds to the event observed later, $\bar{C}$ (non-C) the opposite case. Now if the clairvoyant should have the faculty supposed, the series of events of the type C and $\bar{C}$ would define a series with a limit of the frequency. If the man should be a perfect prophet, this limit would be the number 1; however we may admit less perfect prophets with a lower limit. Anyway, we have constructed here a series with a limit. We must have such a series if we want to control the prophet; our control

[22] Cf. *ibid.*, p. 36.

would consist in nothing but the application of the principle of induction to the series of events C and $\bar{C}$, i.e., in an inductive inference as to the reliability of the prophet, based on his successes. Only if the reduction to such a series with a limit is possible can we know whether or not the man is a good prophet because only this reduction gives us the means of control.

We see from this consideration that the case imagined is not more general but less general than our world of reducible series. A forecast giving us a true determination of every event is a much more special case than the indication of the limit of the frequency and is therefore included in our inductive procedure. We see, at the same time, that our postulate of the existence of limits of the frequencies is not a restriction of the concept of predictability. Any method of prediction defines by itself a series with a limit of the frequency; therefore, if prediction is possible, there are series with limits of the frequencies.

We are entitled, therefore, to call the applicability of the inductive procedure a necessary condition of predictability. We see at the same time why such a relation holds: *it is a logical consequence of the definition of predictability.* This is why we can give our demonstration of the unique position of the inductive principle by means of tautological relations only. *Although the inductive inference is not a tautology, the proof that it leads to the best posit is based on tautologies only.* The formal conception of logic was placed, by the problem of induction, before the paradox that an inference which leads to something new is to be justified within a conception of logic which allows only empty, i.e., tautological, transformations: this paradox is solved by the recognition that the "something new" furnished by the inference is not maintained as a true statement but as our best posit, and that the demonstration is not directed toward the truth of the conclusion but to the logical relation of the procedure to the aim of knowledge.

There might be raised, instinctively, an objection against our theory of induction: that there appears some thing like "a necessary condition of knowledge"—a concept which is accompanied since Kant's theory of knowledge by rather an unpleasant flavor. In our theory, however, this quality of the inductive principle does not spring from any a priori qualities of human reason but has its origin in other sources. He who wants something must say what he wants; he who wants to predict must say what he understands by predicting. It we try to find a definition of this term which corresponds, at least to some extent, to the usual practice of language, the definition—independently of further determination—will turn out to entail the postulate of the existence of certain series having a limit of the frequency. It is from this component of the definition that the character of the inductive principle as being a necessary condition of predictability is deduced. The application of the principle of induction does not signify, therefore, any restriction or any renunciation of predictability in another form—it signifies nothing but the mathematical interpretation of what we mean by predictability, properly speaking.

We turn now to a second objection. It was the claim of the first objection that our definition of predictability demands too much; the second objection, on the contrary, holds that this definition demands too little, that what we call predictability is not a sufficient condition of actual predictions. This objection arises from the fact that our definition admits infinite series of events; to this conception is opposed the view that a series actually observable is always finite, of even a rather restricted length, determined by the short duration of human lives.

We shall not deny the latter fact. We must admit that there may be a series of events having a limit whose convergence begins so late that the small portion of the series observed by human beings does not reveal any indication of the later convergence. Such a series would have for us the character of a nonconverging series. Applying the principle of induction, we should never have success with our inferences; after a short time, our posits would always turn out false. Although, in such a case, the condition of predictability would be fulfilled, the inductive procedure would not be a practically sufficient means for discovering it.

We shall not deny this consequence either. We do not admit, however, that the case considered raises any objection to our theory. We did not start for our justification of induction from a presupposition that there are series having a limit; in spite of this, we contrived to give the justification sought. This was made possible by the use of the concept of necessary condition; we said that, if we are not sure of the possibility of success, we should at least realize its necessary conditions. The case of convergence coming too late amounts to the same thing as the case of nonconvergence, as far as human abilities are concerned. However, if we succeed in giving a justification of the inductive procedure even if this worst of all cases cannot be excluded a priori, our justification will also have taken account of the other case—the case of a convergence which is too late.

Let us introduce the term *practical limit* for a series showing a sufficient convergence within a domain accessible to human observations; we may add that we may cover by this term the case of a series which, though not converging at infinity, shows an approximate convergence in a segment of the series, accessible in practice and sufficiently long (a so-called "semiconvergent series"). We may then say that our theory is not concerned with a mathematical limit but with a practical limit. Predictability is to be defined by means of the practical limit, and the inductive procedure is a sufficient condition of success only if the series in question has a practical limit. With these concepts, however, we may carry through our argument just as well. The applicability of the inductive procedure may be shown, even within the domain of these concepts, to be the necessary condition of predictability.

It is the concept of necessary condition on which our reasoning is based. It is true that, if the series in question should have no practical limit—including the case of too late a convergence—this would imply the inefficiency of the inductive procedure. The possibility of this case, however, need not restrain us from at least wagering on success. Only if we knew that the unfavorable case is actual, should we renounce attempts at prediction. But obviously this is not our situation. We do not know whether we shall have success; but we do not know the contrary either. Hume believed that a justification of induction could not be given because *we do not know whether we shall have success;* the correct fomulation, instead, would read that a justification of induction could not be given if *we knew that we should have no success*. We are not in the latter situation but in the former; the question of success is for us indeterminate, and we may therefore at least dare a wager. The wager, however, should not be arbitrarily laid but chosen as favorably as possible; we should at least actualize the necessary conditions of success, if the sufficient conditions are not within our reach. The applicability of the inductive procedure being a necessary condition of predictability, this procedure will determine our best wager.

We may compare our situation to that of a man who

wants to fish in an unexplored part of the sea. There is no one to tell him whether or not there are fish in this place. Shall he cast his net? Well, if he wants to fish in that place I should advise him to cast the net, to take the chance at least. It is preferable to try even in uncertainty than not to try and be certain of getting nothing.

§ 41. Concatenated inductions

The considerations concerning the possibility of too slow a convergence of the series could not shake our justification of the inductive procedure, as signifying at least an attempt to find a practically convergent series; they do point out, however, the utility of methods which would lead to a quicker approximation, i.e., which would indicate the true value of the limit at a point in the series where the relative frequency is still rather different from the limiting value. We may want even more; we may want methods which give us the numerical value of the limit before the physical actualization of the series has begun—a problem which may be considered as an extreme case of the first problem. The elaboration of such methods is indeed a question of the greatest relevance; we shall ask now whether or not they exist, and how they are to be found.

We have already met with an example which may be considered as the transition to a method of quicker approximation. We discussed the possibility of a clairvoyant and said that his capacities might be controlled by the inductive principle; we said that, should the control confirm the predictions, the clairvoyant was to be considered a reliable prophet, and his indications as superior to those of the inductive principle. This idea shows an important feature of inductive methods. We may sometimes infer by means of the inductive principle that it is better to apply some other method of prediction; the inductive principle may lead to its own supersession. This is no contradiction; on the contrary, there is no logical difficulty in such a procedure—it even signifies one of the most useful methods of scientific inquiry.

If we want to study inferences of such a type, we need not trouble clairvoyants or oracles of a mystic kind: science itself has developed such methods to a vast extent. The method of scientific inquiry may be considered as a concatenation of inductive inferences, with the aim of superseding the inductive principle in all those cases in which it would lead to a false result, or in which it would lead us too late to the right result. It is to this procedure of concatenated inductions that the overwhelming success of scientific method is due. The complication of the procedure has become the reason why it has been misinterpreted by many philosophers; the apparent contradiction to a direct application of the inductive principle, in individual cases, has been considered as a proof for the existence of noninductive methods which were to be superior to the "primitive" method of induction. Thus the principle of causal connection has been conceived as a noninductive method which was to furnish us with an "inner connection" of the phenomena instead of the "mere succession" furnished by induction. Such interpretations reveal a profound misunderstanding of the methods of science. There is no difference between causal and inductive laws; the former are nothing but a special case of the latter. They are the case of a limit equal to 1, or at least approximately equal to 1; if we know, in such a case, the value of the limit even before the series has begun, we have the case of the individual prediction of future events happening in novel conditions, such as is demanded within the causal conception of knowledge. This case, therefore, is included in our theory of concatenated inductions.

The connecting link, within all chains of inferences leading to predictions, is always the inductive inference. This is because among all scientific inferences there is only one of an overreaching type: that is the inductive inference. All other inferences are empty, tautological; they do not add anything new to the experiences from which they start. The inductive inference does; that is why it is the elementary form of the method of scientific discovery. However, it is the only form; there are no cases of connections of phenomena assumed by science which do not fit into the inductive scheme. We need only construct this scheme in a sufficiently general form to include all methods of science. For this purpose, we must turn now to an analysis of concatenated inductions.

We begin with a rather simple case which already shows the logical structure by which the inductive inference may be superseded in an individual case. Chemists have found that almost all substances will melt if they are sufficiently heated; only carbon has not been liquefied. Chemists do not believe, however, that carbon is infusible; they are convinced that at a higher temperature carbon will also melt and that it is due only to the imperfection of our technical means that a sufficiently high temperature has not yet been attained. To construe the logical structure of the inferences connected with these experiences, let us denote by A the melted state of the substance, by $\bar{A}$ the contrary state, and arrange the states in a series of ascending temperatures; we then have the scheme

Copper:	$\bar{A}\ \bar{A}\ A\ A\ A\ A\ A\ A\ A \ldots$
Iron:	$\bar{A}\ \bar{A}\ \bar{A}\ A\ A\ A\ A\ A\ A \ldots$
———	. . . .
———	. . . .
Carbon:	$\bar{A}\ \bar{A}\ \bar{A}\ \bar{A}\ \bar{A}\ \bar{A}\ \bar{A}\ \bar{A}\ \bar{A} \ldots$

To this scheme, which we call a *probability lattice*, we apply the inductive inference in two directions. The first is the horizontal. For the first lines it furnishes the result that above a certain temperature the substance will always be in liquid state. (Our example is a special case of the inductive inference, where the limit of the frequency is equal to 1.) For the last line, the corresponding inference would furnish the result that carbon is infusible. Here, however, an inference in the vertical direction intervenes; it states that in all the other cases the series leads to melting, and infers from this that the same will hold for the last line if the experiment is sufficiently continued. We see that here a cross-induction concerning a series of series occurs, and that this induction of the second level supersedes an induction of the first level.

This procedure may be interpreted in the following way. Applying the inductive principle in the horizontal direction, we proceed to posits concerning the limit of the frequency; these are blind posits, as we do not know a co-ordinated weight. Presupposing the validity of these posits, we then count in the vertical direction and find that the value 1 has a high relative frequency among the horizontal limits, whereas the value 0 furnished by the last line is an exception. In this way we obtain a weight for the horizontal limits; thus the blind posits are transformed into posits with appraised weight. Regarding the weights obtained we now correct the posit of the last line into one with the highest weight. The procedure may therefore be conceived as a transformation of blind posits into posits with appraised weights, combined with corrections following from the weights obtained—a typical probability method, based on the frequency interpretation. It makes use of the existence of probabilities of different levels. The frequency within the horizontal lines determines a probability of the first level; counting the frequency within a series the elements of which are themselves series we obtain a probability of the second level.[23] The probability of the second level determines the weight of the sentence stating a probability of the first level. We must not forget, however, that the transformation into an appraised posit concerns only the posits of the first level, whereas the posits of the second level remain blind. Thus at the end of the transformation there appears a blind posit of higher level. This of course may also be transformed into a posit with appraised weight, if we incorporate it into a higher manifold, the elements of which are series of series; it is obvious that this transformation will again furnish a new blind posit of a still higher level. We may say: Every blind posit may be transformed into a posit with appraised weight, but the transformation introduces new blind posits. Thus there will always be some blind posits on which the whole concatenation is based.

Our example concerns a special case in so far as the limits occurring are 1 and 0 only. If we want to find examples of the general case, we must pass to cases of statistical laws. To have a model of the inferences occurring, let us consider an example of the theory of games of chance, chosen in such a form that simplified inferences occur.

Let there be a set of three urns containing white and black balls in different ratios of combination; suppose we know that the ratios of the white balls to the total number of balls are 1:4, 2:4, and 3:4, but that we do not know to which urn each of these ratios belongs. We choose an urn, then make four draws from it (always putting the drawn

[23] As to the theory of probabilities of higher levels cf. the author's *Wahrscheinlichkeitslehre*, §§ 56–60.

ball back into the urn before the following draw), and obtain three white balls. Relative to further draws from the same urn, we have now two questions:

1. What is the probability of a white ball?

According to the inductive principle, this question will be answered by 3/4. This is a blind posit. To transform it into a posit with appraised weight, we proceed to the second question:

2. What is the probability that the probability of a white ball is 3/4?

This question concerns a probability of the second level; it is equivalent to the question as to the probability judged on the basis of the draws already made that the chosen urn contains the ratio 3/4. The calculus of probability, by means of considerations also involving a problem of a probability lattice, gives to this question a rather complicated answer which we need not here analyze; in our example it furnishes the value 27/46. We see that though our best posit in the given case will be the limit 3/4 of the frequency, this posit is not very good; it itself has only the weight 27/46. Considering the next drawing, as a single case, we have here two weights: the weight 3/4 for the drawing of a white ball, and the weight 27/46 for the value 3/4 of the first weight. The second weight in this case is smaller than the first; if, to obtain a comparison, we write the weights in decimal fractions, we have 0.75 for the first and 0.59 for the second weight.

In this example the original posit is confirmed by the determination of the weight of the second level, this one being greater than 1/2, and therefore greater than the second level weight belonging to the wagers on the limit 2/4 or 1/4. By another choice of the numerical values, a case of correction would result, i.e., a case in which the weight of the second level would incline us to change the first posit. If there were twenty urns, nineteen of which contained white balls in a ratio of 1/4, and only one contained white balls in the ratio of 3/4, the probability at the second level would become 9/28 = 0.32; in such a case, we should correct the first posit and posit the limit 1/4, in opposition to the principle of induction. The occurrence of three white balls among four would then be regarded as a chance exception which could not be considered as a sufficient basis for an inductive inference; this correction would be due to the change of a blind posit into an appraised one.

Our example is, as we said, simplified; this simplification is contained in the following two points. First, we presupposed some knowledge about the possible values of the probabilities of the first level: that there are in dispute only the three values 1/4, 2/4, and 3/4 (in the second case: only the two values 1/4 and 3/4). Second, we presupposed that the urns are equally probable for our choice, i.e., we attribute to the urns the initial probabilities 1/3 (in the second case: 1/20); this presupposition is also contained in the calculation of the value 27/46 (in the second case: 9/28) for the probability of the second level.

In general, we are not entitled to such presuppositions. We are rather obliged to make inquiries as to the possible values of the probabilities of the first level and their corresponding initial probabilities. The structure of these inferences is also to be expressed in a probability lattice, but of a type more general than that used in the example concerning the melting of chemical substances; the limits of the frequencies occurring here are not just 1 or 0. The answers can only be given in the form of posits based on frequency observations, so that the whole calculation involves still further posits and posits of the blind type. This is why we cannot dispense with blind posits; although each can be transformed into an appraised posit, new blind posits are introduced by the transformation itself.[24]

Before we enter into an analysis of this process leading to posits and weights of higher levels, we must discuss some objections against our probability interpretation of scientific inferences. It might be alleged that not all scientific inferences are purely of the probability type and so not fully covered by our inductive schema. The objection may run that there are causal assumptions behind our inferences without which we should not venture to place our wagers. In our chemical example, the posit of the limit 1 in the horizontal lines of the figure is not only based on a simple enumeration of the A's and $\bar{A}$'s; we know that if a substance is once melted it will not become solid at a higher temperature. Neither is our positing the possibility of liquefying carbon at higher temperatures based on simply counting the lines of the figure; we know from the atomic theory of matter that heat, in increasing the velocity of the atoms, must have the effect of decomposing the structure of the solids. Causal assumptions of this kind play a decisive role in such inferences as furnished by the example.

Although we shall not deny the relevance of considerations of this kind as far as the actual inference of the physicist is concerned, their occurrence, however, does not preclude the possibility that these so-called causal assumptions admit an interpretation of the inductive type. We simplified our analysis to show the inductive structure of the main inferences; what is shown by the objection is that an isolation of some of the inductive chains is not correct, that every case is incorporated in the whole concatenation of knowledge. Our thesis that all inferences occurring are

[24] For an exact analysis of these inferences cf. the author's *Wahrscheinlichkeitslehre*, § 77.

of the inductive type is not thereby shaken. We shall show this by another example which will clarify the inductive nature of so-called causal explanations.

Newton's law of gravitation has always been considered as the prototype of an explicative law. Galileo's law of falling bodies and Kepler's law of the elliptic motion of celestial bodies were inductive generalizations of observed facts; but Newton's law, it is said, was a causal explication of the facts observed. Newton did not observe facts but reflected upon them; his idea of an attractive force explained the motions observed, and the mathematical form he gave to his ideas shows no resemblance to methods of probability such as occur in our scheme. Is not this a proof against our inductive interpretation of scientific inferences?

I cannot admit this. On the contrary, Newton's discovery seems to me to involve typical methods of the probability procedure of science. To show this, let us enter into a more detailed analysis of the example.

The experiments of Galileo were performed on falling bodies whose spatiotemporal positions he observed; he found that the quantities measured fit into the formula $s = gt^2/2$, and inferred, by means of the inductive principle, that the same law holds for similar cases. Let us denote by A the case that the spatiotemporal values measured fulfil the relation $s = gt^2/2$; we have then a series in which A has been observed with a relative frequency almost equal to 1, and for which we maintain a limit of the frequency at 1. Correspondingly, Kepler observed a series of spatiotemporal positions of the planet Mars and found that they may be connected by a mathematical relation which he called the Law of Areas. If we again denote by A the case that the relation is fulfilled by the spatiotemporal values, we also obtain a series in which A has a relative frequency of almost 1, and for which a limit at 1 is inferred. The contrary cases $\bar{A}$ (non-A) include those cases, never wholly to be eliminated, in which the observations do not fit into the mathematical relation. As the observations of both examples relate to not one but numerous series of experiments, we have to represent them by the following schema:

$$\left.\begin{array}{ccccccccc} A & A & A & A & A & \bar{A} & A & A & \ldots \\ . & . & . & . & . & . & . & . & \\ . & . & . & . & . & . & . & . & \end{array}\right\} \text{Galileo}$$

$$\left.\begin{array}{ccccccccc} A & A & A & \bar{A} & A & A & A & A & \ldots \\ . & . & . & . & . & . & . & . & \\ . & . & . & . & . & . & . & . & \end{array}\right\} \text{Kepler}$$

It is the discovery of Newton that a formula may be given which includes the observations of both Galileo and Kepler; we may therefore consider the preceding scheme, consisting of two parts, as one undivided scheme for which the case A is defined by one mathematical relation only. It is the famous relation $k(m_1 m_2/r^2)$ which does this; the case A may be regarded as meaning the correspondence of observations to this mathematical law, in both parts of the scheme.

With this recognition, the applicability of probability methods is greatly expanded. We are now able to apply cross-inductions leading from the Galilean lines of the scheme to the Keplerian lines, and inversely; i.e., the validity of Kepler's laws is no longer based on Kepler's observational material alone but jointly on Galileo's material, and conversely, the validity of Galileo's law is jointly supported by Kepler's observational material. Before Newton, similar cross-inductions were only possible within each section of the schema separately. Newton's discovery, therefore, in unifying both theories, involves an increase of certainty for both of them; it links a more comprehensive body of observational material together to form one inductive group.

The increase of certainty described corresponds to the conception of the men of science shown on the occasion of theoretical discoveries of this kind. Classical logic and epistemology could not assign any valid argument for this interpretation; it is only probability logic which, by the idea of concatenated inductions, is able to justify such a conception. We see that only in placing the causal structure of knowledge within the framework of probability do we arrive at an understanding of its essential features.

§ 42. The two kinds of simplicity

It might be objected to our interpretation that, logically speaking, Newton's discovery is trivial; if a finite set of observations of very different kinds is given, it is always mathematically possible to construct a formula which simultaneously embraces all the observations. In general such a formula would be very complicated, even so complicated that a human mind would not be able to discover it; it is the advantage of Newton's discovery that in this case a very simple formula suffices. But this, the objection continues, is all Newton did; Newton's theory is simpler, more elegant than others—but progress in the direction of truth is not connected with his discovery. Simplicity is a matter of scientific taste, a postulate of scientific economy, but has no relation to truth.

This kind of reasoning, well known from many a positivistic writer, is the outcome of a profound misunderstanding of the probability character of scientific methods. It is true that for any set of observations a comprehensive formula may be constructed, at least theoretically, and that Newton's formula is distinguished by simplicity from all the others. But this simplicity is not a matter of scientific taste; it has on the contrary an inductive function, i.e., it brings to Newton's formula good predictional qualities. To show this, we must add a remark concerning simplicity.

There are cases in which the simplicity of a theory is nothing but a matter of taste or of economy. These are cases in which the theories compared are logically equivalent, i.e., correspond in all observable facts. A well-known case of this type is the difference of systems of measurement. The metrical system is simpler than the system of yards and inches, but there is no difference in their truth-character; to any indication within the metrical system there is a corresponding indication within the system of yards and inches—if one is true, the other is true also, and conversely. The greater simplicity in this case is really a matter of taste and economy. Calculations within the metrical system permit the application of the rules of decimal fractions; this is indeed a great practical advantage which makes the introduction of the metrical system desirable in those countries which still keep to the yard and inch system—but this is the only difference. For this kind of simplicity which concerns only the description and not the facts co-ordinated to the description, I have proposed the name *descriptive simplicity*. It plays a great role in modern physics in all those places where a choice between definitions is open to us. This is the case in many of Einstein's theorems; it is the reason the theory of relativity offers a great many examples of descriptive simplicity. Thus the choice of a system of reference which is to be called the *system in rest* is a matter of descriptive simplicity; it is one of the results of Einstein's ideas that we have to speak here of descriptive simplicity, that there is no difference of truth-character such as Copernicus believed. The question of the definition of simultaneity or of the choice of Euclidean or non-Euclidean geometry are also of this type. In all these cases it is a matter of convenience only for which definition we decide.

However, there are other cases in which simplicity determines a choice between nonequivalent theories. Such cases occur when a diagram is to be drawn which is determined by some physical measurements. Imagine that a physicist found by experiment the points indicated on Figure 6; he wants to draw a curve which passes through

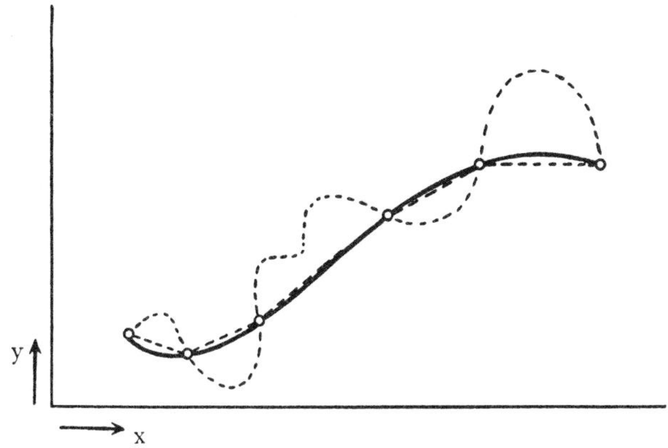

FIG. 6.—The simplest curve: inductive simplicity

the data observed. It is well known that the physicist chooses the simplest curve; this is not to be regarded as a matter of convenience. We have drawn in Figure 6, in addition to the simplest curve, one (the dotted line) which makes many oscillations between the observed points. The two curves correspond as to the measurements observed, but they differ as to future measurements; hence they signify different predictions based on the same observational material. The choice of the simplest curve, consequently, depends on an inductive assumption: we believe that the simplest curve gives the best predictions. In such a case we speak of *inductive simplicity;* this concept applies to theories which differ in respect to predictions, although they are based on the same observational material. Or, more precisely speaking: The relation "difference as to inductive simplicity" holds between theories which are equivalent in respect to all observed facts, but which are not equivalent in respect to predictions.[25]

The confusion of both kinds of simplicity has caused much mischief in the field of the philosophy of science. Positivists like Mach have talked of a principle of economy which is to replace the aim of truth, supposedly followed by science; there is, they say, no scientific truth but only a most economical description. This is nothing but a confusion of the two concepts of simplicity. The principle of economy determines the choice between theories which differ in respect to descriptive simplicity; this idea has been erroneously transferred to cases of inductive simplicity, with the result that no truth is left at all but only economy. Actually in cases of inductive simplicity it is not economy which determines our choice. The regulative principle of the construction of scientific theories is the postulate of the best predictive character; all our decisions as to the choice between unequivalent theories are determined by this postulate. If in such cases the question of simplicity plays a certain role for our decision, it is because we make the assumption that the simplest theory furnishes the best predictions. This assumption cannot be justified by convenience; it has a truth-character and demands a justification within the theory of probability and induction.

Our theory of induction enables us to give this justifica-

[25] The terms "descriptive simplicity" and "inductive simplicity" have been introduced in the author's *Axiomatik der relativistischen Raum-Zeit-Lehre* (Braunschweig, 1924), p. 9. A further elucidation of these concepts has been given in the author's *Ziele und Wege der physikalischen Erkenntnis* in *Handbuch der Physik,* ed. Geiger-Scheel (Berlin, 1929), IV, 34–36.

tion. We justified the inductive inference by showing that it corresponds to a procedure the continued application of which must lead to success, if success is possible at all. The same idea holds for the principle of the simplest curve.

What we want to construct with the diagram is a continuous function determining both past and future observations, a mathematical law of the phenomena. Keeping this aim before our eyes, we may give a justification of the procedure of the simplest curve, by dividing our reasoning into two steps.

In the first step, let us imagine that we join the observed points by a chain of straight lines, such as drawn in Figure 6. This must be a first approximation; for if there is a function such as we wish to construct, it must be possible to approximate it by a chain of straight lines. It may be that future observation will show too much deviation; then, we shall correct our diagram by drawing a new chain of straight lines, including the newly observed points. This procedure of preliminary drawing and later correction must lead to the true curve, if there is such a curve at all—its applicability is a necessary condition of the existence of a law determining the phenomena.

It is the method of anticipation which is adopted with such a procedure. We do not know whether our observed points are sufficiently dense to admit a linear approximation to the curve; but we anticipate this case, being ready to correct our posit if later observations do not confirm it. At some time we shall have success with this procedure—if success is attainable at all.

But the chain of straight lines does not correspond to the actual procedure applied by the physicist. He prefers a smooth curve, without angles, to the chain of straight lines. The justification of this procedure necessitates a second step in our considerations.

For this purpose we must consider the derivatives of the function represented by the curve. The differential quotients of a function are regarded in physics as physical entities, in the same sense as is the original entity represented by the function; thus, if the original entity is a spatial distance represented as a function of time, the first derivative is a velocity, the second an acceleration, etc. For all these derived entities we aim also to construct mathematical laws; we want to find for them also continuous functions such as are sought in our diagram. Regarding the chain of straight lines from this point of view, it already fails for the first derivative; in this case the first differential quotient, designed as a function of the argument x, is not represented by a continuous curve but by a discontinuous chain of horizontal lines. This may be illustrated by Figure 7, the dotted lines of which correspond to the first derivative of the chain of straight lines of Figure 6; we see that we do not obtain here even a continuous chain of straight lines but a chain broken up into several parts. Thus, if we approximate the original curve by a chain of straight lines, the principle of linear approximation is followed only as to the original curve; for the first derivative it is already violated. This is different, however, for the smooth curve; its derivatives, conceived as functions of x, are smooth curves as well. This may be seen in Figure 7, where the first derivative of the smooth curve of Figure 6 is represented by the continuous smooth line. This is the reason for the preference of the smooth curve. It has, in respect to the set of observed points, qualities similar to those of a linear interpolation and may be justified by the principle of anticipation as well; moreover, it also satisfies the same postulate for its derivatives.

The procedure of the smoothest interpolation may be considered, therefore, as a superposition of linear interpolations carried through for the construction of the original function and of its derivatives. Thus the nonlinear interpolation by the smoothest curve may be justified by a reduction to linear interpolations which determine, on the whole, a nonlinear interpolation to be preferable. The procedure corresponds not to a single induction but to a concatenation of inductions concerning different functions standing in the mutual relation of a function and its derivative; the result is a better induction, as it is based on a repeated application of the inductive principle, and incorporates corrections in the sense defined in § 41.

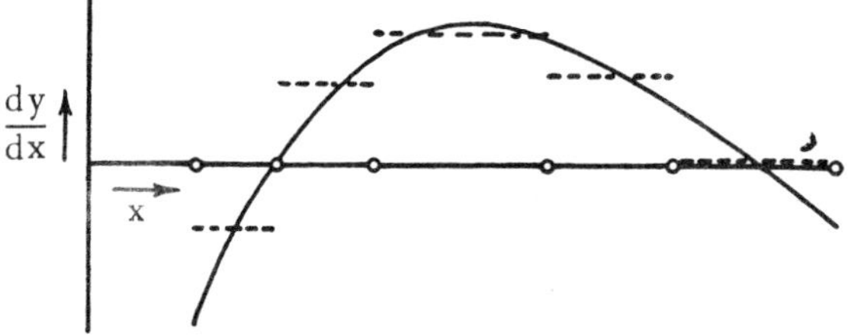

FIG. 7.—Derivatives of the simplest curve, and of the chain of straight lines, developed from Fig. 6.

There remains an objection to our reasoning. We contrived to justify the preference of the smooth curve to the chain of straight lines; but the postulate of the smooth curve is not unambiguous. Though a curve such as drawn in the dotted line of Figure 6 is excluded, there remain other smooth curves very similar to the one drawn; the points observed will not furnish us a clear decision as to the choice between such similar smooth curves. Which are we to choose?

Here we must answer that the choice is not relevant. From the viewpoint of approximation, there is no great difference as to these forms of curves; all of them converge asymptotically; they do not differ essentially as far as predictions are concerned. The choice between them may therefore be determined from the viewpoint of convenience. The principle of inductive simplicity determines the choice only to a certain extent: it excludes the oscillating curve drawn in Figure 6, but there remains a small domain of indeterminacy within which the principle of descriptive simplicity may be applied. We prefer here a simpler analytical expression because we know better how to handle it in a mathematical context; this is permissible because the functions open to our choice do not relevantly differ as to predictions of further observations between the points observed.

To the latter argument a further objection may be raised. It is true that within the domain of the observed points there is no great difference between all these smooth curves; but this is no longer valid outside this domain. All analytical functions define a prolongation of the curve into a distant domain, and two analytical functions which differ only slightly within the interior domain, may lead to great differences as to extrapolations. Consequently the choice between them can not be justified by descriptive simplicity as far as extrapolations are concerned; how, then, may we justify this choice?

To this we must answer that a set of observations does not at all justify an extrapolation of any considerable length. The desire to know the continuation of the curve far beyond the observed domain may be very strong with the physicist; but, if he has nothing but the observed set at his disposal, he must renounce any hypothesis concerning extrapolations. The inductive principle is the only rule the physicist has at hand; if it does not apply, philosophy cannot provide him with a mysterious principle showing the way where induction fails—in such a case, there remains nothing but to confess a modest *ignoramus*.

Our adversary might object that the man of science does not always comply with this alternative. Only the spirit of mediocrity will submit to renunciation, he will exclaim; the scientific genius does not feel bound to the narrow restrictions of induction—he will guess the law outside the domain of observed facts, even if your principle of induction cannot justify his presentiments. Your theory of induction as an interpolation, as a method of continual approximation by means of anticipations, may be good enough for the subordinate problems of scientific inquiry, for the completion and consolidation of scientific theories. Let us leave this task to the artisans of scientific inquiry—the genius follows other ways, unknown to us, unjustifiable a priori, but justified afterward by the success of his predictions. Is not the discovery of Newton the work of a genius which never would have been achieved by methods of simple induction? Is not Einstein's discovery of new laws of the motion of planets, of the bending of light by gravitation, of the identity of mass and energy, etc., a construction of ideas which has no relation to diagrams of curves of interpolation, to statistics of relative frequencies, to the slow driving of approximations, step by step?

Let me say that I should be the last to discredit the work of the great men of science. I know as well as others that the working of their minds cannot be replaced by directions for use of diagrams and statistics. I shall not venture any description of the ways of thought followed by them in the moments of their great discoveries; the obscurity of the birth of great ideas will never be satisfactorily cleared up by psychological investigation. I do not admit, however, that these facts constitute any objection against my theory of induction as the only means for an expansion of knowledge.

We pointed out in the beginning of our inquiry (§ 1) the

distinction between the context of discovery and the context of justification. We emphasized that epistemology cannot be concerned with the first but only with the latter; we showed that the analysis of science is not directed toward actual thinking processes but toward the rational reconstruction of knowledge. It is this determination of the task of epistemology which we must remember if we want to construct a theory of scientific research.

What we wish to point out with our theory of induction is the logical relation of the new theory to the known facts. We do not insist that the discovery of the new theory is performed by a reflection of a kind similar to our exposition; we do not maintain anything about the question of how it is performed—what we maintain is nothing but a relation of a theory to facts, independent of the man who found the theory. There must be some definite relation of this kind, or there would be nothing to be discovered by the man of science. Why was Einstein's theory of gravitation a great discovery, even before it was confirmed by astronomical observations? Because Einstein saw—as his predecessors had not seen—that the known facts indicate such a theory; i.e., that an inductive expansion of the known facts leads to the new theory. This is just what distinguishes the great scientific discoverer from a clairvoyant. The latter wants to foresee the future without making use of induction; his forecast is a construction in open space, without any bridge to the solid domain of observation, and it is a mere matter of chance whether his predictions will or will not be confirmed. The man of science constructs his forecast in such a way that known facts support it by inductive relations; that is why we trust his prediction. What makes the greatness of his work is that he sees the inductive relations between different elements in the system of knowledge where other people did not see them;

but it is not true that he predicts phenomena which have no inductive relations at all to known facts. Scientific genius does not manifest itself in contemptuously neglecting inductive methods; on the contrary, it shows its supremacy over inferior ways of thought by better handling, by more cleverly using the methods of induction, which always will remain the genuine methods of scientific discovery.

That there is an inductive relation from the known facts to the new theory becomes obvious by the following reflection. The adherents of the contrary opinion believe that the construction of the new theory is due to a kind of mystic presentiment but that later, after a confirmation of the predictions contained in the new theory, it is proved to be true. This is, however, nothing but one of the unwarranted schematizations of two-valued logic. We shall never have a definitive proof of the theory; the so-called confirmation consists in the demonstration of some facts which confer a higher probability upon the theory, i.e., which allow rather simple inductive inferences to the theory. The situation before the confirmation differs from that after it only in degree. This situation is characterized by the occurrence of some facts which confer at least some probability upon the theory and which distinguish it from others as our best posit, according to inductive methods. This is what the good theorist sees. If there were no such inductive relations, his supposition would be a mere guess, and his success due to chance only.

We may add the remark that the distinction of the context of justification from the context of discovery is not restricted to inductive thinking alone. The same distinction applies to deductive operations of thought. If we are faced by a mathematical problem, say, the construction of a triangle from three given parameters, the solution (or the class of solutions) is entirely determined by the given problem. If any solution is presented to us, we may decide unambiguously and with the use of deductive operations alone whether or not it is correct. The way in which we find the solution, however, remains to a great extent in the unexplored darkness of productive thought and may be influenced by aesthetic considerations, or a "feeling of geometrical harmony." From the reports of great mathematicians it is known that aesthetic considerations may play a decisive role in their discoveries of great mathematical theorems. Yet in spite of this psychological fact, no one would propound a philosophical theory that the solution of mathematical problems is determined by aesthetic points of view. The objective relation from the given entities to the solution and the subjective way of finding it are clearly separated for problems of a deductive character; we must learn to make the same distinction for the problem of the inductive relation from facts to the theories.

There are cases, it is true, in which a clear decision as to the most favorable theory cannot be obtained because there are several theories with equal weights indicated by the facts. This does not mean that we are at a loss with the inductive principle; on the contrary, a great number of theories is always ruled out by this principle. But among the weights of the admissible remainder there may be no maximum, or so flat a maximum that it cannot be considered as furnishing the basis for a clear decision. In such cases, which we may call cases of *differential decision*,[26] different men of science will decide for different

[26] I choose this name by analogy with the term "differential diagnosis" used by physicians, to denote a case where the observed symptoms of illness indicate several diseases as their possible origin but do not permit a decision among the members of this group unless certain new symptoms can be observed. This differential diagnosis is, logically speaking, a special case of our differential decision.

theories, their decisions being determined by personal taste more than by scientific principles; the final decision will then be made by later experiments of a crucial character. It is a kind of "natural selection," of "struggle for existence," which determines in such a case the final acceptance of a scientific theory; though this case happens, and not too rarely, we must not forget that this is just a case in which scientific prophecy breaks down, the decision in favor of an assumption being possible only after the occurrence of the predicted events. The man who predicted the right theory is then sometimes considered a great prophet because he knew the true prediction even in a case when scientific principles of prediction failed. But we must not forget that his success is the success of a gambler who is proud, having foreseen the *rouge* or the *noir*. This presumed prophetic gift will always expose its spurious nature in a second case of prediction when success will be wanting. A man of science, in the case of differential decision, had better admit that he cannot rationally make his choice.

In the context of our introduction of the concept of inductive simplicity, we illustrated its meaning by a diagram and pointed out a smooth curve as the model of this kind of simplicity. However, this is not the only case of this kind. The inductive connections of modern physics are constructed analytically; this is why the theorist of physics must be a good mathematician.[27]

The inductive procedure of Newton consisted in his demonstration that a simple mathematical formula covers both Galileo's and Kepler's laws. The simplicity of the formula expresses its character as an interpolation, as a linear,

[27] We may add that the graphical interpretation of inductive inferences may be also carried through, for complicated cases, if we pass to a parameter space of a higher number of dimensions (cf. the author's article, "Die Kausalbehauptung und die Möglichkeit ihrer empirischen Nachprüfung," *Erkenntnis*, III [1932], 32).

or almost linear, approximation; it is this quality to which its predictional qualities are due. Newton's theory not only incorporates the observations of Galileo and Kepler but also leads to predictions. The "predictions" may concern phenomena which are already known, but which were neither seen before in connection with the other phenomena nor used as a part of the basis on which the new theory was constructed. Of such a kind was Newton's explanation of the tides. On the other hand, Newton's theory led also to predictions, properly speaking, e.g., the attraction of a ball of lead to other bodies such as observed by Cavendish in the turning of a torsion balance.

We raised the question whether in a diagram an extrapolation is possible which extends to a domain rather far removed from the domain of the points observed. There are examples in which extrapolations of such a kind seem to occur. Such cases, however, are to be otherwise explained; there are facts of another type, not belonging to the domain of the observation points marked in the diagram, which support the extrapolation. Examples of this kind are cases in which the analytical form of the curve is known to the physicist before the observations, and these are made only to determine the numerical constants of the analytical expression. This case, which happens rather frequently in physics, corresponds in our example to a determination of the curve by facts outside the observed domain; for the analytical form of the curve is then determined by reflections connecting the phenomenon in question to other phenomena.

An example of a similar type is Einstein's prediction of the deviation of light rays emitted by stars in the gravitational field of the sun. Had he pursued only the plan of finding a generalization of Newton's law of planetary motion such that the irregularities of the planet Mercury would have been explained, his hypothesis of the deviation of light would have been an unwarranted extrapolation not justified by inductions. But Einstein saw that a much more comprehensive body of observations was at his disposal, which could be interpolated by means of the idea that a gravitational field and an accelerated motion are always equivalent. From this "equivalence principle" the deviation of light rays followed immediately; thus within the wider context Einstein's prediction was the "smoothest interpolation." It is this quality which is denoted in predicates frequently applied to Einstein's theories, such as "the natural simplicity of his assumptions"; such predicates express the inductive simplicity of a theory, i.e., its character of being a smooth interpolation. This does not diminish the greatness of Einstein's discovery; on the contrary, it is just his having seen this relation which distinguishes him from a clairvoyant and makes him one of the most admirable prophets within the frame of scientific methods. The gift of seeing lines of smooth interpolation within a vast domain of observational facts is a rare gift of fate; let us be glad we have men who are able to perform in respect to the whole domain of knowledge inferences whose structure reappears in the modest inferences which the artisan of science applies in his everyday work.

§ 43. The probability structure of knowledge

Our discussion of the methods of scientific research and of the formation of scientific theories has led us to the result that the structure of scientific inferences is to be conceived as a concatenation of inductive inferences. The elementary structure of the concatenation is the probability lattice; we may refer here to the exposition of this form of inference in § 41. As a consequence of idealizations, in which the transition from probability to practical truth plays a decisive role, the probability character of the inferences is not always easily seen; the short steps of inductive inferences can be combined into long chains forming longer steps of so complicated a structure that it may be difficult to see the inductive inference as the only atomic element in them. To indicate the method of decomposition of such structures, and of their reduction to inductive inferences, we may here add a discussion of some examples.

There are cases in which one experiment may decide the fate of a theory. Such cases of an *experimentum crucis* are often quoted against the inductive conception of science; they seem to prove that it is not the number of instances which decides in favor of a theory but something such as an "immediate insight into the very nature of the phenomenon," opened for us by one single experiment. On a closer consideration, the procedure is revealed as a special case of concatenated inductions. We may know from previous experience that only two possibilities are left for a certain experiment, i.e., we may know, with great probability, that *A* will be followed by *B* or by *C* and, besides, that there is a great probability that *A* will always be followed by the same type of event, not alternately by both. In such a case, if the probabilities occurring are high, one experiment may indeed suffice for the decision. Of such a type was Lavoisier's decisive experiment concerning combustion. There were in practice only two theories left as an explanation of combustion: the first maintained that a specific substance, phlogiston, escaped during the combustion; the second assumed that a substance originating from air entered the burning body during the combustion. Lavoisier showed in a famous experiment that the body was heavier after being burnt than before; thus one experiment could decide in favor of the oxidation theory of combustion. Yet this was possible only because former inductions had excluded all but two theories and because former inductions had made it very probable that all processes of combustion are of the same type. Thus the *experimentum crucis* finds its explanation in the theory of induction and does not involve further assumptions; it is only the superimposition of a great many elementary inductive inferences which creates logical structures whose form as a whole, if we cling to a schematized conception, suggests the idea of noninductive inference.

It is the great merit of John Stuart Mill to have pointed out that all empirical inferences are reducible to the *inductio per enumerationem simplicem.* The exact proof, however, has been achieved only by the demonstration that the calculus of probability can be reduced to this principle, a demonstration which presupposes an axiomatic construction of the calculus of probability. Physics applies in its inferences, besides logic and mathematics in general, the methods of the calculus of probability; thus an analysis of the latter discipline was as necessary for epistemology as an analysis of logic and the general methods of mathematics.

It is on account of this foundation of probability inferences on the principle of induction that we are entitled to interpret the inferences leading from observations to facts as inductive inferences. Inferences appearing in the form of the schemas developed within the calculus of probability are reducible, for this reason, to inductive inferences. Of this kind are many inferences which, on superficial examination, show no probability character at all but look like a decision concerning an assumption, based on an observation of its "necessary consequences." If a detective infers from some fingerprints on a bloody knife that Mr. X is a murderer, this is usually justified by saying: It is impossible that another man should have the same fingerprints as Mr. X; it is impossible that the bloody knife lying beside the dead body of the victim was not used to kill the man, under the given conditions, and so on. These so-called impossibilities are, however, only very low probabilities, and the whole inference must be considered as falling under the rule of Bayes, one of the well-known schemas of the calculus of probability which is used for inferring from given observations the probabilities of their causes. It furnishes, consequently, not a certainty but only a high probability for the assumption in question.

Scientific inferences from observations to facts are of the same type. If Darwin maintained the theory that the logical order of organisms according to the differentiation of their internal structure, may be interpreted as the historical order of the development of the species, this theory is based on facts such as the correspondence of the time order of geological layers (determined by their lying one above the other) to the occurrence of higher organisms. With the assumption of a theory which considers the higher organisms as old as the lowest ones, this correspondence would appear as a very improbable result. Conversely, according to Bayes's rule the observed fact makes Darwin's theory probable and the other theory improbable. The probability character of this inference is usually veiled by the use of statements such as, "The other theory is incompatible with the observed facts," a statement in which the transition from a low probability to impossibility is performed; and epistemological conceptions have been developed according to which a theory is unambiguously tested by its consequences. A trained eye nevertheless discovers probability structures in all these inferences from facts to theories. With this analysis the reduction of the inferences occurring to inductive inferences is also performed, owing to the reducibility of the calculus of probability to the inductive principle. This is the reason why we may say that scientific inferences from facts to theories are inductive inferences.

Scientific induction is not of a form "higher" than the ordinary inductions of daily life; but it is better in the sense of a difference of degree. This difference is due to the concatenation of inductions such as expressed in the application of the rules of the calculus of probability; they lead to results which by direct inductions would never be attained. We said that the inductive nature of these inferences is sometimes obscured by a schematization in which probability implications are replaced by strict implications; this may be illustrated by another example. Some philosophers have distinguished a generalizing from an exact induction; the first is to be our poor frequency-bound induction, which is restricted to probabilities only; whereas the second is to be a higher method of cognition which, though based on experience, is to lead to absolute certainty. I may refer here to a discussion I once had with a biologist of high rank, who refused to admit that his science is dependent on so imperfect a principle as *inductio per enumerationem simplicem*. He presented to me an example concerning carnivorous and herbivorous animals. We observe, he argued, that the first have a short intestine, the latter a long one; we infer then by generalizing induction that there is a causal connection between the food and the length of the intestine. This is only a mere supposition, he said; yet it is proved later by exact induction at the moment we succeed in experimentally changing the length of the intestine by the food we give to the animal. Such experiments have indeed been successfully performed with tadpoles.[28] But what is overlooked in such reasoning is

[28] Cf. Max Hartmann, "Die methodologischen Grundlagen der Biologie," *Erkenntnis*, III (1932–33), 248.

that the difference in question is nothing but a difference of degree. The experiments with tadpoles enlarge the observational material, and precisely in a direction which permits us to make use of certain laws well established by previous inductions, such as the law that food has an influence on the development of the organism, that the other conditions in which the animals were kept do not influence in general their intestines, and the like. I do not say this to depreciate the work of biologists; on the contrary, the progress of knowledge from lower probabilities to higher ones is due to experiments of such a kind. There is no reason though to construct a qualitative difference of methods where quantitative differences are in question. What the experimental scientist does is to construct conditions in which all of the processes occurring except the one which is to be tested are conformable to known cases; by this isolation of the unknown phenomenon from other unknown phenomena he arrives at simpler forms of the inductive inference. As to the interpretation of this procedure we must take care not to confound an idealization with the inferences actually occurring. If we consider those high probabilities occurring as equal to 1, we transfrom the actual procedure into a schema in which "causal connections" occur, and in which one experiment may demonstrate with certainty some new "causal law." To infer from the applicability of such a scheme the existence of an "exact induction" which is to be of a logical type different from the ordinary induction, means overstraining an approximation and drawing conclusions which are valid for the schema only and not for the real procedure to which it applies.

Any epistemology which forces knowledge into the frame of two-valued logic is exposed to this danger. It was the grave mistake of traditional epistemology to consider knowledge as a system of two-valued propositions; it is to this conception that all kinds of apriorism are due, these being nothing but an attempt to justify an absolutely certain knowledge of synthetic character. And it is to this conception also that all kinds of skepticism are due, renunciation of truth being the attitude of more critical minds before the problem of such absolute knowledge. The way between Scylla and Charybdis is pointed out by the probability theory of knowledge. There is neither an absolutely certain knowledge nor an absolute ignorance—there is a way between them pointed out by the principle of induction as our best guide.

If we say that the two-valued logic does not apply to actual knowledge, this is not to maintain that it is false. It is to affirm only that the conditions of its application are not realized. Scientific propositions are not used as two-valued entities but as entities having a weight within a continuous scale; hence the presuppositions of two-valued logic are not realized in science. Treating science as a system of two-valued propositions is like playing chess on a board whose squares are smaller than the feet of the pieces; the rules of the game cannot be applied in such a case because it remains indeterminate on which square a piece stands. Similarly, the rules of two-valued logic cannot be applied to scientific propositions, at least not generally, because there is no determinate truth-value corresponding to the propositions, but only a weight. It is therefore probability logic alone which applies to knowledge in its general structure.

Is there no way, we may be asked, to escape this consequence? Is there no way of transforming probability logic into the two-valued logic? As to the answer to this question, we may make use of our inquiries concerning this transformation (§ 36). We showed that there are two ways for making such a transformation. The first one is the way of dichotomy or trichotomy; we found that this way can only lead to an approximative validity of the two-valued logic. The second way makes use of the frequency interpretation; however it is also restricted to an approximative validity for two reasons: first, because the individual element of the propositional series is not strictly true or false and, second, because the frequency to be asserted cannot be asserted with certainty. It is this latter point of view which we must now analyze more accurately.

The transition under consideration can be conceived, if we use the logical conception of probability (§ 33), as a transition from probability statements to statements about the probability of other statements; but it would be erroneous to believe that in this way we could arrive at a strict logic of two values. A statement about the probability of another statement is in itself not true or false but is only given to us with a determinate weight. Using the transition in question, we shall never arrive at something other than probabilities. We are bound to this flight of steps leading from one probability into another. It is only a schematization if we stop at one of the steps and regard the high probability obtained there as truth. It was a schematization, therefore, when we spoke throughout our inquiry of *the* predicate of weight; we should have spoken of an infinite set of weights of all levels co-ordinated to a statement. We may refer here to our numerical example (§ 41) in which we calculated the probability 0.75 for a statement of the first level and the probability 0.59 for the statement of the second level that the first statement has the probability 0.75; in this example, we cut off the flight at the second step. This was also a schematization, owing to the simplified conditions in which the problem was given; an exhaustive consideration would have to take into account all probabilities of the infinitely numerous levels.

From the example given we also see another feature of the probability structure of knowledge: that the probabilities occurring are by no means all either of a high or of a low degree. There are intermediate degrees as well; their calculation may be based on the frequency of elementary propositions whose probability is near to the extreme values 1 or 0 (cf. § 36)—but the propositions to which these probabilities are co-ordinated as weights enter into the system of knowledge as propositions of an intermediate degree of weight. This is why for the whole of science two-valued logic does not even apply in the sense of an approximation. An approximative application of two-valued logic obtains only if we consider not the direct propositions of science but those of the second or a higher level—propositions about the probability of direct propositions of science.[29]

The occurrence of different probabilities of higher levels is a specific feature of probability logic; two-valued logic shows this feature only in a degenerate form. Our probability of the second level would correspond in the two-valued logic to the truth of the sentence, "The sentence *a* is true"; but if *a* is true, then "*a* is true" is true also. Thus we need not consider the truth-values of higher levels in the two-valued logic; this is why this problem plays no role in traditional logic or logistic. In probability logic, on the other hand, we cannot dispense with considerations of this

[29] In our preceding inquiries we frequently made use of the approximate validity of two-valued logic for the second-level language. One schematization of this kind is that we considered statements about the weight of a proposition as being true or false; another one is contained in our use of the concepts of physical and logical possibility, occurring in our definitions of meaning. Strictly speaking, there is, between these types of possibility, a difference of degree only. We were entitled to consider them in a schematized form as qualitatively different because they concern reflections belonging to the second-level language. The approximate validity of two-valued logic for the second-level language also explains why the positivistic language can be conceived as approximately valid in the sense of a second-level language (cf. the remark at the end of § 17).

sort; that is why the application of probability logic to the logical structure of science is a rather complicated matter.

These reflections become relevant if we want to define the probability of a scientific theory. This question has attained some significance in the recent discussion of the probability theory of knowledge. The attempt has been made to show that probability logic is not a sufficiently wide framework to include scientific theories as a whole. Only for simple propositions, it has been said, may a probability be determined; for scientific theories we do not know a definite probability, and we cannot determine it because there are no methods defining a way for such a determination.

This objection originates from underrating the significance of the probabilities of higher levels. We said it was already a schematization if we spoke of *the* probability, or *the* weight of a simple proposition; this schematization, however, is permissible as a sufficient approximation. But this no longer holds if we pass from simple propositions to scientific theories. For example, there is no such thing as *the* probability of the quantum theory. A physical theory is a rather complex aggregate; its different components may have different probabilities which should be determined separately. The probabilities occurring here are not all of the same level. To a scientific theory belongs, consequently, a set of probabilities, including probabilities of the different parts of the theory and of different levels.[30]

Within the analysis of the problem of the probability of theories, one question in particular has stood in the foreground of discussion. It has been asked whether the probability of a theory concerns the facts predicted by the

[30] These different probabilities cannot in general be mathematically combined into one probability; such a simplification presupposes special mathematical conditions which would apply, if at all, only to parts of the theory (cf. *Wahrscheinlichkeitslehre*, § 58).

theory, or whether we have to consider the theory as a sociological phenomenon and to count the number of successful theories produced by mankind. The answer is that both kinds of calculation apply but that they correspond to different levels. The quantum theory predicts a great many phenomena, such as observations on electrometers and light rays, with determinate probabilities; as the theory is to be considered as the logical conjunction of propositions about these phenomena, its probability may be determined as the arithmetical product of these elementary probabilities. This is the probability of the first level belonging to the quantum theory. On the other hand, we may consider the quantum theory as an element in the manifold of theories produced by physicists and ask for the ratio of successful theories within this manifold. The probability obtained in this way is to be interpreted not as the direct probability of the quantum theory but as the probability of the assumption, "The quantum theory is true"; as the truth occurring here is not strict truth but only a high probability, namely, that of the first level, the probability of the second level is independent of that of the first and demands a calculation of its own. We see that at least two probabilities of different levels play a role in questions about theories; we might construct still more, considering other kinds of classification of the theory. If we add a consideration of the fact that the parts of a theory may already belong to different levels, we see that a theory within the probability theory of knowledge is not characterized by a simple weight but by a set of weights partially comprising weights of the same, partially of different, levels.

The practical calculation of the probability of a theory involves difficulties, but it would be erroneous to assume that our conception lacks any practical basis. It is true that the probability of theories of a high generality is usually not quantitatively calculated; but as soon as determinations of a numerical character occur within science, such as those concerning physical constants, they are combined with calculations which may be interpreted as preliminary steps toward the calculation of the probability of a theory. It is the application of the mathematical theory of errors in which considerations of this kind find their expression. The "average error" of a determination may be interpreted, according to well-known results of the calculus of probability,[31] as the limits within which the deviation of future observations will remain with the probability 2/3; thus this indication may be conceived as the calculation of a first-level probability of an assumption. If we say that "the velocity of light is 299796 km/sec, with an average error of ± 4, or of ± 0.0015 per cent"[32] this may be read: "The probability that the velocity of light lies between 299792 km/sec and 299800 km/sec, is 2/3." It can easily be shown that we may infer from this a lower limit for the probability (on the first level) of Einstein's hypothesis of the constancy of the velocity of light; passing to somewhat wider limits of precision and applying some properties of the Gaussian law, we may state this result in this form: "The probability of Einstein's hypothesis of the constancy of the velocity of light is greater than 99.99 per cent, if a numerical range of 0.0052 per cent is admitted for the possible value of the constant." Considerations of a similar type may be carried through for theories of a more comprehensive character.

As to probabilities of the second level, we cannot as yet determine their numerical values. It has been objected that we here meet a difficulty of principle because we do

[31] Cf. *ibid.*, p. 226.

[32] A. A. Michelson, *Astrophysical Journal*, LXV (1927), 1.

not know into which class the theory is to be incorporated if we want to determine its probability in the frequency sense; thus if we want to determine the second-level probability of the quantum theory, shall we consider the class of scientific theories in general, or only that of physical theories, or only that of physical theories in modern times? I do not think that this is a serious difficulty, as the same question occurs for the determination of the probability of single events; I have indicated in § 34 the method of procedure in such a case. The narrowest class available is the best; it must, however, be large enough to afford reliable statistics. If the probability of theories (of the second level) is not yet accessible to a quantitative determination, the reason is to be found, I think, in the fact that we have in this field no sufficiently large statistics of uniform cases. That is to say, if we use a class of cases not too small in number, we may easily indicate a subclass in which the probability is considerably different. We know this from general considerations, and thus we do not try to make statistics. Future statistics may perhaps overcome these difficulties, as the similar difficulties of meteorological statistics have been overcome. As long as we have no such statistics, crude appraisals will be used in their place—as in all fields of human knowledge not yet accessible to satisfactory quantitative determinations. Appraisals of this kind (concerning the second-level probability of a theory) may acquire practical importance in cases when we judge a theory by the success obtained with other theories in that domain; if an astronomer propounds a new theory of the evolution of the universe, we hesitate to trust this theory on account of unfortunate experiences with other theories of that kind.

A last objection remains. We said that a theory, and even a simple proposition, is characterized not by a single weight but by a set of weights infinite in number. We must in any case confine ourselves to a finite number of members. This would be justified if all the following members should be weights of the degree 1; we might then consider the last weight used as truly determined. But, if we know nothing about all the rest of the set, how can we omit all of them? How can we justify using the weights of the lower levels if we do not know anything about the weights of the higher levels?

To see the force of this objection, let us imagine the case that all the rest of the weights are of a very low degree—near zero. This would result in the last weight determined by us being unreliable; the preceding weight would consequently become unreliable as well, and, as this unreliability is equally transferred to the weight of the first level, the whole system of weights would be worthless. How can we justify our theory of weights, and with this the probability procedure of knowledge, before the irrefutable possibility of such a case?

This objection is nothing but the well-known objection to which the procedure of induction is already exposed in its simplest form. We do not know whether we shall have success in laying our wager corresponding to the principle of induction. But we found that, as long as we do not know the contrary, it is advisable to wager—to take our chance at least. We know that the principle of induction determines our best wager, or posit, because this is the only posit of which we know that it must lead to success if success is attainable at all. As to the system of concatenated inductions, we know more: we know that it is better than any single induction. The system, as a whole, will lead to success earlier than a single induction; and it may lead to success even if some single inductions should remain without success. This logical difference, the superiority of the net of concatenated inductions to single inductions, can be demonstrated by purely mathematical considerations, i.e., by means of tautologies; hence our preference for the system of inductions can be justified without any appeal to presuppositions concerning nature. It is very remarkable that such a demonstration can be given; although we do not know whether our means of prediction will have any success, yet we can establish an order between them and distinguish one of them, the system of concatenated inductions, as the best. With this result the application of the system of scientific inductions finds a justification similar to, and even better than, that of the single induction: *the system of scientific inductions is the best posit we know concerning the future.*

We found that the posits of the highest level are always blind posits; thus the system of knowledge, as a whole, is a blind posit. Posits of the lower levels have appraised weights; but their serviceableness depends on the unknown weights of the posits of higher levels. The uncertainty of knowledge as a whole therefore penetrates to the simplest posits we can make—those concerning the events of daily life. Such a result seems unavoidable for any theory of prediction. We have no certainty as to foreseeing the future. We do not know whether the predictions of complicated theories, such as the quantum theory or the theory of albumen molecules, will turn out to be true; we do not even know whether the simplest posits concerning our immediate future will be confirmed, whether they concern the sun's rising or the persistence of the conditions of our personal environment. There is no principle of philosophy to warrant the reliability of such predictions; that is our answer to all attempts made within the history of philosophy to procure for us such certainty, from Plato, through all varieties of theology, to Descartes and Kant.

In spite of that, we do not renounce prediction; the arguments of skeptics like Hume cannot shake our resolution: at least to *try* predictions. We know with certainty that among all procedures for foreseeing the future, known to us as involving success if success is possible, the procedure of concatenated inductions is the best. We try it as our best posit in order to have our chance—if we do not succeed, well, then our trial was in vain.

Is this to say that we are to renounce any belief in success? There is such a belief; everyone has it when he makes inductions; does our solution of the inductive problem oblige us to dissuade him from this firm belief?

This is not a philosophical but a social question. As philosophers we know that such a belief is not justifiable; as sociologists we may be glad that there is such a belief. Not everyone is likely to act according to a principle if he does not believe in success; thus belief may guide him when the postulates of logic turn out to be too weak to direct him.

Yet our admission of this belief is not the attitude of the skeptic who, not knowing a solution of his own, permits everyone to believe what he wants. We may admit the belief because we know that it will determine the same actions that logical analysis would determine. Though we cannot justify the belief, we can justify the logical structure of the inference to which it fortunately corresponds as far as the practical results are concerned. This happy coincidence is certainly to be explained by Darwin's idea of selection; those animals were to survive whose habits of belief corresponded to the most useful instrument for foreseeing the future. There is no reason to dissuade anybody from doing with belief something which he ought to do in the same way if he had no belief.

This remark does not merely apply to the belief in induction as such. There are other kinds of belief which have crystallized round the methods of expanding knowledge. Men of scientific research are not always of so clear an insight into philosophical problems as logical analysis would require: they have filled up the world of research work with mystic concepts; they talk of "instinctive presentiments," of "natural hypotheses," and one of the best among them told me once that he found his great theories because he was convinced of the harmony of nature. If we were to analyze the discoveries of these men, we would find that their way of proceeding corresponds in a surprisingly high degree to the rules of the principle of induction, applied however to a domain of facts where average minds did not see their traces. In such cases, inductive operations are imbedded within a belief which as to its intension differs from the inductive principle, although its function within the system of operations of knowledge amounts to the same. The mysticism of scientific discovery is nothing but a superstructure of images and wishes; the supporting structure below is determined by the inductive principle.

I do not say this with the intention to discredit the belief —to pull the superstructure down. On the contrary, it seems to be a psychological law that discoveries need a kind of mythology; just as the inductive inference may lead us in certain cases to the preference of methods different from it, it may lead us also to the psychological law that sometimes those men will be best in making inductions who believe they possess other guides. The philosopher should not be astonished at this.

This does not mean that I should advise him to share any of these kinds of belief. It is the philosopher's aim to know what he does; to understand thought operations and not merely to apply them instinctively, automatically. He wants to look through the superstructure and to discover the supporting structure. Belief in induction, belief in a uniformity of the world, belief in a mystic harmony between nature and reason—they belong, all of them, to the superstructure; the solid foundation below is the system of inductive operations. The difficulty of a logical justification of these operations misled philosophers to seek a justification of the superstructure, to attempt an ontological justification of inductive belief by looking for necessary qualities of the world which would insure the success of inductive inferences. All such attempts will fail—because we shall never be able to give a cogent proof of any material presumption concerning nature. The way toward an understanding of the step from experience to prediction lies in the logical sphere; to find it we have to free ourselves from one deep-rooted prejudice: from the presupposition that the system of knowledge is to be a system of true propositions. If we cross out this assumption within the theory of knowledge, the difficulties dissolve, and with them dissolves the mystical mist lying above the research methods of science. We shall then interpret knowledge as a system of posits, or wagers; with this the question of justification assumes as its form the question whether scientific knowledge is our best wager. Logical analysis shows that this demonstration can be given, that the inductive procedure of science is distinguished from other methods of prediction as leading to the most favorable posits. Thus we wager on the predictions of science and wager on the predictions of practical wisdom: we wager on the sun's rising tomorrow, we wager that food will nourish us tomorrow, we wager that our feet will carry us tomorrow. Our stake is not low; all our personal existence, our life itself, is at stake. To confess ignorance in the face of the future is the tragic duty of all scientific philosophy; but, if we are excluded from knowing true predictions, we shall be glad that at least we know the road toward our best wagers.

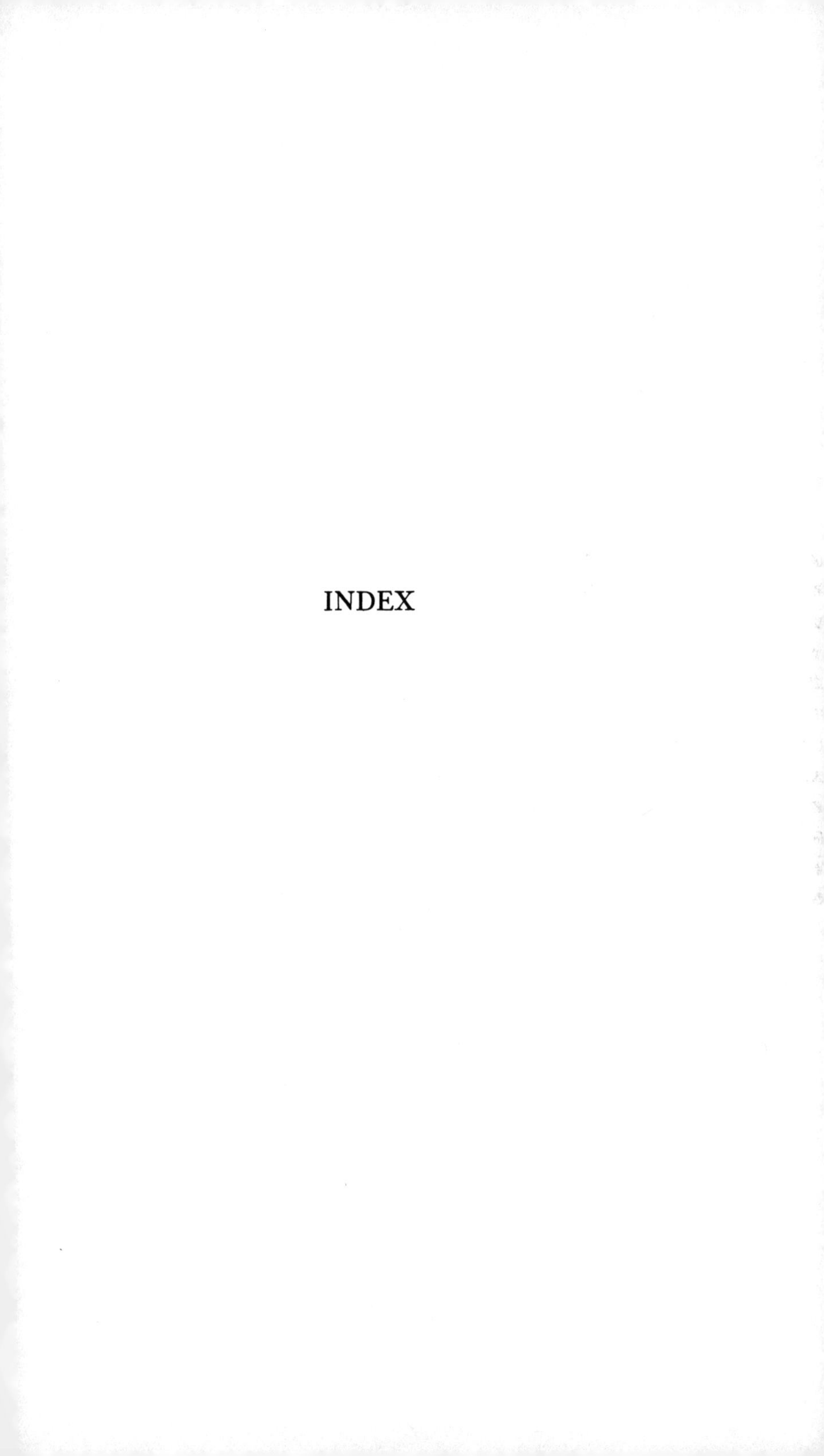

INDEX

INDEX

Abstracta, 93, 211, 235; existence of, 93, 101
Action: and meaning, 70, 80, 309, 344; and weight, 25, 32, 64, 315
Analysis of science, 8
Anticipation, method of, 353, 377
Aristotle, 299, 347
Atom, 215, 263, 267; existence of, 213
Avenarius, Richard, 163
Average error, 398

Bacon, 341, 347
Bases of epistemological construction, 203, 262
Basic statement, 173; in the narrower sense, 181; in the wider sense, 181
Basis: atom, 215, 263, 267; concreta, 263, 275; impression, 263; internal-process, 263; proposition, 263, 268; reaction, 263; stimulus, 263
Bayes, rule of, 124
Behaviorism, 163, 240
Bodily feeling, 235, 259
Bohr, 157
Boltzmann, L., 213
Boole, 299, 334, 342
Bühler, 60

Carnap, R., 5, 38, 60, 76, 145, 163, 171, 204, 226, 269, 335, 338
Causality, 317, 364, 370, 373, 392; homogeneity of, 139
Cavendish, 386
Class; *see* Probability
Complex, 98, 105; disjunctive, 107, 111; existence of the, 111; projective, 111, 130; reducible, 111, 130, 143
Composition, relation of, 99
Concatenation of inductive inferences, 363, 387, 391
Concreta, 93, 98, 210, 214, 265
Constancy of the velocity of light, 398
Constitutive relations, 107
Context: of discovery, 7, 382; of justification, 7, 382
Convention, 9
Conventionalism, 14, 271
Copeland, 298
Cross-induction, 366, 372

Darwin, 390
Decision, 146; differential, 384; entailed, 13, 63
Deducibility relation, 269, 336
Demarcation value, 327
De Morgan, 299
Descartes, 85, 261, 334, 344, 401
Description, 196
Dewey, 49, 163
Discovery, context of, 7, 382
Disparity conception, 300, 302, 325, 338
Dörge, 298
Dream, 92, 102, 139, 144, 165, 202, 205

Economy, principle of, 376
Ego, 152, 259
Einstein, 9, 43, 78, 127, 381, 386
Elements: complete set of, 107; external, 110; internal, 98, 105, 110
Equally possible cases, 300
Equally probable, 305
Euclidean geometry, 12, 271
Evidence, 285

Existence: of abstracta, 93, 101; grammar of the word, 195; of external things, 90, 102, 111, 129, 133; immediate, 199, 218; independent, 115, 132; objective, 199, 204, 218; reducibility of, 105, 114; return to the basis of immediate, 204, 275; subjective, 199, 204
Existential coupling, 201
Experimentum crucis, 388

Fact: logical, 11, 336; object, 11; physical, 83; single, 84
Falsification, 88
Fermat, 298
Fleck, L., 224
Frequency interpretation, 300, 304, 329, 337
Freud, 208, 246

Galileo, 371, 385
Gauss, 298
Gestalt, 100, 221

Hartmann, M., 391
Helmholtz, 9
Hempel, C. G., 37
Hertz, P., 357
Hilbert, 335
Hume, David, 73, 78, 262, 335, 341, 348, 356, 362, 401

Identity conception, 300, 302, 325, 338
Illata, 212, 227
Implication, strict, 269
Impression, 88, 171; and external things, 101, 115, 129, 132, 144; and form, 173; stereoscopic, 228
Inductio per enumerationem simplicem, 389
Induction, 339, 356, 382, 400; aim of, 350; belief in, 402; concatenation of, 363, 387, 391; cross-, 366, 372; formulation of the principle of, 340; justification of the principle of, 348; as necessary condition, 349
Inductive inferences, concatenation of, 387
Inner process, 226
Insufficient reason, principle of, 306
Interactional quality, 168
Intuitiveness, 178
Introspection, 227, 234
Isomorphism of the two probability concepts, 303

James, W., 49
Justification, context of, 7, 382

Kant, 234, 334, 346, 401
Kantianism, 12
Kepler, 371, 385
Keynes, J. M., 300, 302, 332
Kokoszynska, Marja, 37
Kolmogoroff, 298

Language, 16, 57; analysis of, 270; of chess, 28; communicative function of, 59; egocentric, 135, 140, 147; emotional function of, 60; impression 149, 237; inner-process, 237; objective, 266; reaction, 232; relaxive function of, 60; second-level, 155; stimulus, 231; subjective, 266; suggestive function of, 59; syntax of, 336
Laplace, 298, 301
Lavoisier, 388
Leibnitz, 78, 299, 334
Lévy-Bruhl, 205
Lewis, C. I., 151, 269
Lichtenberg, 261
Limit, practical, 361
Localization: of abstracta, 98; of external objects, 223; of psychical phenomena, 234; within our body, 167
Locke, 164
Löwy, H., 261
Logic: alternative, 321, 326; aprioristic conception of, 334; formalistic conception of, 334, 343, 359; probability, 319, 326, 336, 373;

of propositional series, 325; two-valued, 321, 326; of weights, 324
Lukasiewicz, 300

Mach, 78, 213, 376
Meaning, 17, 20, 30, 59, 63, 80; and action, 70; functional conception of, 156; logical, 40, 55, 62, 124, 134; physical, 40, 55, 72; physical-truth, 55, 62, 127, 149; probability, 54, 62, 124, 127, 149, 153, 160; probability theory of, 54, 71, 87, 133, 189; super-empirical, 62, 68; truth, 55, 148; truth theory of, 30, 37, 53, 101, 191; verifiability theory of, 55, 57, 77, 79, 95, 189, 305
Memory, reliability of, 180
Michelson, A. A., 84, 398
Mill, John Stuart, 342, 389
Mises, v., 298
Modality, 320
Mysticism, religious, 58

Necessary condition, 349, 351, 354, 356, 360
Necessity, 320, 336
Neurath, 163
Newton, 78, 371, 373, 385
Nominalism, 93

Ogden, 60
Ontology, 98, 336, 404
Overreaching character, 127, 130, 132, 365

Parallels, convergence of, 223
Part and whole, 99
Pascal, 298
Passivity in observation, 258
Peirce, C. S., 49, 299, 339
Perceptual function, constancy of the, 183
Plato, 334, 401
Poincaré, 14
Popper, K., 88, 302, 338
Posit, 313, 352, 366; appraised, 352; blind, 353, 366
Positivism, 30, 79, 156, 265; and existence, 101, 112, 129; and meaning, 30, 72, 189; as a problem of language, 145
Possibility, 38, 320
Pragmatism, 30, 48, 57, 69, 79, 150, 163
Predictability, 350, 359
Predictional value, 26, 190, 315
Predictions, 339, 360, 381, 401; of a clairvoyant, 353, 358; included in every statement, 85, 131
Presentation, 89
Probability, 24, 75, 292, 297; a posteriori, 340; a priori, 124; axiomatic of, 337; backward, 124; class determining the degree of, 316, 399; concatenation of, 274; connection, 52, 104, 109, 192, 244; of disjunction, 173; forward, 124; frequency interpretation of, 300, 304, 329, 337; of higher level, 331, 395; implication 51; inference, 130, 142; initial, 124, 369; lattice, 366; logic, 319, 326, 336, 373; logical concept of, 299, 303; mathematical concept of, 298; of a scientific theory, 396; of the second level, 367, 368; series with changing probabilities, 310; series, order of, 317; of the single case, 302, 305, 309, 352
Projection, 110, 129, 136, 143, 212; internal, 216
Proposition, 20; basic, 173; co-ordination of, 95, 108; direct, 47; impression, 89, 169; indirect, 47; molecular, 21; observation, 34, 37, 87; predicates of, 19, 188; religious, 65; report, 276, 282
Propositional series, 324, 329
Psychical experience, incomparability of, 248
Psychoanalysis, 208, 246, 287
Psychology, 225

Quale of the sensation, 250
Quantum mechanics, 139, 187, 317

Rational belief, 334, 338
Rational reconstruction, 5
Reaction, 226
Realism, 93, 145, 159
Recollection image, 179
Reducibility relation, 99, 114
Reducible series, 358
Reduction, relation of, 98, 105, 114 136, 148
Relativity of motion, 44
Representation, 209
Retrogression, principle of, 49, 101, 130, 148, 343
Russell, Bertrand, 95, 335

Schaxel, I., 250
Schematization of the conception of knowledge, 104, 157, 188, 383, 393
Schiller, 49
Secondary quality, 167
Self-observation, 234
Semiconvergent series, 361
Sensation, 89
Sense, 20; inner tactile, 237; internal, 164, 226; of touch, 166
Sense data, 89
Sentence, 21
Significant, 158
Similarity: disjunction, 172; immediate, 171
Simplest curve, 375, 379
Simplicity, 373; descriptive, 374; inductive, 376
Simultaneity, 43, 127, 137
Single case; *see* Probability
Specimens, collection of, 35, 182, 202
Statement, 21; *see also* Proposition
Stimulus, 226; *see also* Basis
Subjectivism, 290
Substitute world, 220
Subvocal speaking, 343
Sufficient conditions, 355
Superiority of the immediate present, 281
Symbols, 17

Tarski, A., 37
Task: advisory, 13; critical, 7; descriptive, 3
Tautology, 335
Things: immediate, 199, 289; objective, 199, 276, 289; peremptory character of immediate, 275, 290; subjective, 199, 276
Tolman, E. C., 163
Tornier, 298
Transport time, 128
Trichotomy, 327
Truth, 31, 28, 190; objective, subjective, and immediate, 280, 287; physical theory of, 32, 33
Truth-value, 21, 321

Utilizability, 69, 150, 344

Venn, 300, 334, 342
Verifiability, 30, 38, 304; absolute, 83, 125, 187
Verification, indirect, 46
Volitional bifurcation, 10, 147
Volitional decision, 9
Voltaire, 262

Wager, 314
Watson, 163, 243
Weight, 24, 188, 297, 314, 394; and action, 32, 71; appraisal of, 319, 332, 366, 399; initial, 277; and meaning 75, 120
Weights, system of, 273
Whewell, 342
William of Ockham, 77
Wittgenstein, 49, 74, 335

Zawirski, 300, 321